The Railroad and the Space Program
An Exploration in Historical Analogy

TECHNOLOGY, SPACE, AND SOCIETY

Series Prepared by the American Academy of Arts and Sciences

General Editors:
    Raymond A. Bauer, Harvard Graduate School of
      Business Administration
    Edward E. Furash, Arthur D. Little, Inc.

IN PREPARATION: *Social Indicators,* by Raymond A. Bauer

The Railroad and the Space Program

An Exploration in Historical Analogy

EDITED BY BRUCE MAZLISH

THE M.I.T. PRESS
Massachusetts Institute of Technology
Cambridge, Massachusetts, and London, England

Copyright © 1965 by
The American Academy of Arts and Sciences.

All rights reserved. This book may not be reproduced, in whole or in part, in any form (except by reviewers for the public press), without written permission from the publishers.

Library of Congress Catalog Card Number: 65–28407
Printed in the United States of America

Foreword

This volume is one of a series of publications resulting from an inquiry by the American Academy of Arts and Sciences into the impact of our national space efforts on American society. The research for this inquiry was begun in 1962 at the request of the National Aeronautics and Space Administration. The project was funded under NASA Grant NSg-253-62.

In asking the American Academy of Arts and Sciences to undertake this study, the National Aeronautics and Space Administration funded a wide-ranging and independent probe of the effects of its activities on our society and economy. The American Academy, in turn, has granted the participants in its project the considerable latitude and freedom that is desired in a scholarly investigation. As a result, the published statements from this project reflect the views and findings of its participants rather than the official viewpoints and pronouncements either of the American Academy or of the National Aeronautics and Space Administration.

Although all federal agencies are concerned with the direct consequences of their programs, NASA sponsorship of this inquiry represents a pioneering step. First, NASA has recognized the need to understand not only the closest and most immediate effects of the conquest of space, but also the far-reaching, indirect consequences of the most massive program of technological innovation ever undertaken deliberately. Second, it has been willing to subject the effects of its program to outside scrutiny, without knowing in advance the results. Third, NASA has granted the American Academy the academic freedom which alone can produce a meaningful study.

In this, the first volume to appear in the series *Technology, Space, and Society,* Professor Bruce Mazlish and his associates have focused their attention on historical analogy as a device to assist us in forecasting or projecting the impact of the space program on society. Dr. Mazlish's volume represents one of the first attempts to analyze the theoretical aspects of historical analogy and to determine its usefulness in anticipating the effects of a specific development. The authors chose the coming of the railroad as the analogy most comparable to America's present-day space program. In the nineteenth century, the railroad was a "tremendous and far-reaching engine of social revolution" which left its mark on a whole range of areas—economic, sociological, political, and intellectual—areas which are similar to those being affected by the space program today.

Dr. Mazlish's volume combines a much needed discussion of historical analogy with a revealing study of the American railroad that includes an analysis of the relationship between the railroad and modern technology, the effect of the railroad on the growth of our economy, the railroad's part in creating our modern form of industrial administration, and its effect on people's ideas of the world in which they live. Finally, his study illuminates a number of parallels between the railroad and the space effort, thus providing for NASA and the American people a perceptive guide to the effects of space exploration and their meaning for American society.

EARL P. STEVENSON
Chairman, Committee on Space
American Academy of Arts and Sciences

Preface

The reader ought to be made aware of the circumstances that prompted this book, so that he may come to an informed evaluation of the success or failure of the project involved. As we know, the United States government and people have made a large commitment, morally and financially, to the *primary* purposes of the space effort: the exploration of outer space, the landing of a man on the moon, the surpassing of the Soviet Union in these matters, and so forth. The *secondary*, and even *tertiary*, consequences of the space effort — the economic, social, political, and intellectual impacts, to name a few — are less present to our minds perhaps because less foreseeable. They are not, however, less important. Indeed, in the long run, they may be the most significant aspects of the American space effort, and be so especially for the average citizen in his daily life. Less dramatic than the blasting off of a missile, the secondary consequences promise to be more fundamental, spreading like the tremors of an earthquake through the basic elements of our society.

NASA, in an enlightened and farsighted mood, has tried to give thought to some of the secondary consequences that may result from the fulfillment of its primary mission. As part of this effort of conscience and consciousness, it gave a grant to the American Academy of Arts and Sciences to study the impact of the space effort on society. A Committee on Space was set up, chaired by Mr. Earl Stevenson, with an active working group under the direction of Professor Raymond Bauer of the Harvard Business School. In addition to projects involving the direct study of the space program's effect on such things as technological diffusion, community structure, and manpower (with published volumes on these researches to follow shortly), Professor Bauer wished, boldly, to see whether the study of the past could serve as a "device of anticipation" for the future. Accordingly, he asked that a study in historical analogy be undertaken. To some of the doubts raised he responded by saying that even a negative finding would have positive value.

This, then, is the first problem tackled in the present study. We seek to understand what is meant by historical analogy, as a theoretical problem. Presumably, it is only on this basis that we can judge how useful historical analogy, at least in theory, may be as a "device of anticipation." The discussion of this matter is mainly dealt with in my introduction. Let me anticipate by saying that, to my surprise, I have come to share Professor Bauer's partially sanguine view on the subject.

My optimism rests on the study of *analogous social inventions,* not of bits and pieces of technological or economic data; and I have attempted to indicate what I mean by this on pages 11–18 of my introductory chapter. We chose as our analogous social invention the impact on society of the railroads in the early nineteenth century. Some of the reasons for this choice are spelled out in the introduction: they are further discussed in "A Preliminary Consideration of Historical Analogy" submitted to the Committee on Space, June 28, 1963. I reproduce here some of the arguments presented in that paper:

There are at least three types of analogies to the space effort that are possible. Before I outline these, let me say that one other possibility would be to project ourselves back in time, say, to the 15th–16th century in the age of discoveries or to any period in which any social invention comparable to the space effort was being introduced, and there, so to speak, to set up our committee. Then we could try to imagine what problems such a committee would study in order to find out the secondary consequences of, say, the discovery of America, or the impact of the internal combustion engine. However, such an attempt presents so many difficulties and involves such an immersion in the thinking of the period from which we would make our investigations that it seems precluded as an effort at the present time.

Turning back to our three possible types of analysis, I would suggest the following:

(1), that we can isolate one element in a time and cultural setting different from our own and then study it. For example—and here I borrow from some of the proposals that have already been submitted to the Committee — we can take one element, the microscope, and ask what impact its disclosure of unexpected new life had on the people of the 17th century, or more specifically, on England or Holland, and compare this with the possible impact that the discovery of extraterrestrial life might have on present-day America. We can do this for a whole range of elements. For instance, we might ask similar questions concerning the impact of the internal combustion engine, the discovery of America, and so forth.

My objection to this approach is that it embraces a shot-gun attitude; that is, if we study the impact of the microscope on 17th-century England, we must then establish an enormous number of boundary conditions for that period and that country and compare them with the present conditions. Then, if we make another "element" study, such as the impact of the internal combustion engine, we must establish a whole other set of boundary conditions and compare that to the present. Thus, while the microscope might indicate the impact on imagination, and the internal combustion engine might be studied for its impact on the economy, we have no way of studying the two impacts as one related phenomenon similar or comparable to the impact of the space program on both the imagination and the economy of present-day America.

(2), A second sort of analogy would be to take one element and follow that out over an extended period of time. For example, we could undertake a study of the attitude to space or to the moon held by Western man during the course of the last four or five centuries. Already, there are important, existing studies concerning the changing concepts of time.[1] Undoubtedly, a similar effort concerning man's changing conceptions of space or his attitude to extraterrestrial phenomena, such as the moon, would be important and exciting. However, the actual use of such a study in relation to the space program's impact on society seems rather tangential at this point. Studies such as I have just been talking about would be valuable as a long-range effort, exercises in intellectual and psychological history, trying to establish the major shifts in *Weltanschauungen*.

(3), Lastly, we can try to study the impact of a social invention analogous to that of the space program. In short, a systematic, integrated study of a comparable total impact. After all, the major focus of the present Committee's studies is the impact of the space program in all sorts of areas: technological, economic, sociological, political, intellectual, etc. It would seem, therefore, that the best analogy would be to a social invention, and I use this last in the broad sense of an invention which is, in part, technological (compare, say, the missiles and spacecraft); which is economic in its effect, involving such things as the large-scale employment of manpower, the wide-spread use of materials, and extended, important financial ramifications; which is also political, in the sense of having legislation surrounding its use as well as affecting unification, etc.; and which is sociological in its impact, affecting classes, kinship groups, communities, etc.

Now, clearly, no social invention of the past is going to be absolutely

---

[1] See, for example, Hans Meyerhoff, *Time in Literature* (Berkeley: University of California Press, 1955), and Georges Poulet, *Études sur le Temps Humain* (Paris, 1950).

comparable in all respects to the present social invention. One of the tasks of such a study as proposed here would be to establish the differences as well as the similarities. Thus, our problem is to seize upon a social invention which is most comparable and extends in an important fashion over areas similar to the areas affected by the present space program. The social invention that I recommend is the coming of the railroads.

Pressure of time and money, availability of scholars, and a host of other factors, narrowed this choice to the railroads in America and, even here, to certain segments of the total impact on America; for example, I was unable to arrange either for a detailed study of a railroad community or for a macroscopic survey of the political impact on the country at large.

Nonetheless, the papers presented here — and they form the bulk and value of the volume — constitute what appears to be the first sustained effort to investigate the impact of the railroad on nineteenth-century America in terms of a wide range of inquiries. With all their gaps, these papers form a splendid basis for beginning to understand the multivariate impact of a modern "social invention" on society. They have been written by first-rate scholars, and in many cases they have been based on long-range and extensive studies, some published elsewhere and now focused on this particular problem. Moreover, all of these papers look at hitherto traditional materials, often as much a part of American mythology as of its history, in a manner more modern, cogent, and systematic than usual.

Therefore, the second level on which to judge this volume is as a study of the impact of the railroad on American society. On this level, I have no doubts as to the worth of the book. Alas, it is the problem of comparing the railroad impact to the space impact, analogously, that makes me tremulous, and for good reason. First, of course, an analogy requires at least two parts, and we know very little as yet about the space impact. Second, even what is known about the space impact cannot be expected to form part of the knowledge of practicing historians of the American railroad. The wonder is that some of the contributors have been able to extrapolate at all. I want, however, to delay till a little later on a further discussion of this problem, which involves the question whether a historical study such as the one undertaken here works better by supplying hunches or a systematic model. For the nonce, let us merely agree that the third aspiration of the book is to consider the railroad impact — once studied — as a "device of anticipation" for the study of the space impact.

The fourth aim of this book is really related to what has just been said, but it needs to be remarked on separately. In many ways, it emerged as an unintended consequence of the original task set to us by the Committee on Space. Put simply, it is that the study here undertaken may serve as a prototype for future "impact" studies. No one denies the enormous importance for society of what I have described as "social inventions," yet few rigorous efforts have been made to deal with both the empirical and theoretical aspects of this growing shadow upon all our lives. For this reason, stumbling and faulty as our effort here has been, we can make immodest claims for our inquiry.

Lastly, the present work can be said to have value as an example of the difficulties involved in organizing a project of this kind. In the course of a little over a year, and with a very limited budget in relation to the extraordinary complexity of the problem involved (this is not a complaint, for only a "pilot" project could possibly be justified by NASA), we were asked to say something significant — and back it up concretely — about the possibility of a historical analogy to the impact on society of the space program. In short, the very project itself is a matter for study and judgment in the problems and procedures of the social sciences.

All five problems raised by this book have in common the fact that, to a greater or lesser degree, they have been approached in tentative and exploratory fashion. We have probed rather than proved things. It is on this basis, as a rather novel effort, that the present volume must be judged. Naturally we expect the verdict on each of the five counts to differ, but if we have been successful in casting light on even one of the five problems we shall feel that our journey into space and time has not been without its rewards.

I want now to return to point three, concerning the usefulness of the railroad impact as an analogy to the space impact. Originally, our intention was to draw such analogies in each paper, going, for example, from the economic impact of the railroad to the predictable economic impact of the space program; and in some of the papers, in greater or lesser degree, we have done this. As it turned out, however, such a demand on our historians, involving as it did a knowledge of the space program that was more than superficial, was unfair and unrealistic. As a result, I taxed myself, in the introductory paper, with the task of trying to work out the implications, and hopefully the generalizations, that might legitimately be drawn from the specific studies of the railroad impact for a dimly seen space impact. I hoped that hypotheses or generalizations generated from the example of the railroad would offer us useful tools by which to examine and predict the impact of the space program. Both the introduction and the specific papers on the railroad, as submitted to NASA, are reproduced here with only minor changes, and the reader can judge for himself in this matter.

In this newly written Preface, however, we can now take another look at the problem of analogy as a "device of anticipation." Professor Raymond Bauer, for example, reading over the whole work, has suggested that our primary purpose in drawing an analogy between the railroads and the space program has not been to establish general laws but to secure "insight into the space program." Thus, our "probable inferences" from the railroad are not to be validated in terms of other historical examples in general, but "against the space program." As he sees it, the error in using analogies is to use them "without doing a reasonable job of seeing whether the inference does indeed jibe with available evidence about the current event."

The instance with which Professor Bauer deals probably takes its origin from the assertion by Professors Cootner and Fogel that the maximum impact on the economy by the railroad took place about thirty to fifty years after the innovation of the railroad itself, and from the implied generalization that this is roughly true for all innovations. Professor Bauer comments at some length:

> Let us take a generalization: "The major diffusion of a technological innovation is likely to take place after its period of maximum development when surplus technologists are freed to work in other sectors." The addition of two dozen historical examples, well selected, would increase our confidence in the generality of this statement, but would *not* increase our confidence in its applicability to the space program.
>
> The way to "validate" it for our purpose is something like this:
>
> The first step is to ask whether or not it is sufficiently plausible and important to warrant consideration. Plausibility can be established either by drawing on more historical data, by assessing it intuitively or systematically against our general knowledge, or considering what we know about the space program. Importance is assessed similarly but with increased emphasis on what we know about the space program. By now we have only decided whether or not the proposition is worth investigating. Unfortunately, this is the point at which most people have drawn their conclusions.

We have not yet begun to validate! This must be done in successive stages in each of which we ask two questions: Does it fit the evidence, and is it useful? We first note that there is a lack of correspondence between the two programs in the "half life" of a technologist. Railroad engineers remained reasonably competent throughout their employment period; but the space program makes technologists rapidly obsolete. Therefore, they are likely to become available for other sectors of the economy at an earlier phase of the space program. The proposition did *not* fit this aspect of the evidence, but it was very useful in directing our attention to an important *difference* between the space program and the railroads.

Next, we test the subproposition that diffusion is most likely to take place *via* transfer of people rather than *via* transfer of ideas. Richard Rosenbloom's work [see his forthcoming volume on technological diffusion] gives substance to this part of the proposition. It does not seem to be *absolute,* in that there is some diffusion via transfer of ideas. But the example of the railroads, in turn, reinforces Rosenbloom's findings and suggests serious concern that if nature is let to run its course, the transfer of ideas will not be an adequate substitute for the transfer of people. Here the proposition fits the data quite well and is useful. Other examples can be cited.

His conclusion is that

Analogies may play two roles: the scientific role of developing generalized knowledge and the practical role of illuminating other events. Validation depends on the purpose for which the analogy is drawn. The two purposes are only loosely related to each other. The generality of the proposition must be tested in the way you suggest. This, however, has only a little to do with the power of the analogy to illuminate the other event. This can be tested only by trying it out, and seeing if it does in fact illuminate the other event and in what way.

I find little in Professor Bauer's comment with which to quarrel. Indeed, in my introduction (p. 34), I have remarked that "the highest aim, scientifically, would be the production of generalizations that could then be applied to further materials. This is not to say that hunches or intuitions derived from the study of a historical analogy are to be looked down upon; they are valuable in themselves, and may be all that we can secure." Professor Bauer's "proposition," it seems to me, falls somewhere between my "generalization" and "hunch." Where we disagree is over his *assumption* that the main or decisive factor in the diffusion of a technological innovation is the "surplus technologists . . . freed to work in other sectors." The Cootner-Fogel generalization, on the other hand, suggests that there are multivariate causes operating in terms of the *whole economic structure* of a modern society that prevail *in spite* of the more rapid obsolescence *in one case* of technologists. As Professor Bauer insists, however, verification in terms of the space program itself is the only way to test the legitimacy of the generalization, whose boundary conditions might then have to be drastically redrawn.

Where we differ then, if we do, is in terms of emphasis. Professor Bauer's main concern is in extracting useful and suggestive propositions from the history of the railroad impact that will direct our attention to aspects of the space program that we might otherwise overlook; and I share this concern. On the other hand, I would give greater stress to the usefulness of *attempting* to develop generalized knowledge as, ultimately, the best and surest way of understanding and illuminating specific events, including that of the impact of the space program (a concern I also know Professor Bauer to share). If nothing else, the effort to secure generalizations (which are always transient and time bound at best in the human sciences) forces us to be as systematic and comprehensive as possible. The two ways of proceeding, outlined above, however, are both legitimate and both part of a continuum of "scientific method."

Another consequence of the effort at historical analogy is less pretentious as to "scientific" character. Put simply, it is that the study of the railroad's impact on Americans gives one a "feel" for what is possible in the case of the space impact. For example, many people, especially farmers, can often predict the weather without being able to specify the factors that enter into their decision. Once scoffed at, this sort of "tacit knowledge" is increasingly being granted respect by philosophers. Historians, therefore, can now be more relaxed about their age-old belief that the study of the past, in some vague and ill-defined way, produces a useful sense of what is possible in the future. To say this, however, is not to invoke any relaxation in the effort to secure the most rigorous and systematic knowledge possible of both the past and the future.

In sum, the studies that now follow this Preface offer materials that, at the least specified level, give one a "feel" for the possible impact of what I have called a "social invention." On the next level, they offer suggestions and insights that may helpfully direct our attention to features of a specific impact, that of the space program, which we might otherwise overlook; and, on the most ambitious and specified level, they raise the question as to the possibility and validity of developing generalized knowledge, a "model for the future," concerning the impact on society of various "social inventions." We can justify the boldness of these aims by the reflection that, though it is true that "Fools rush in where angels fear to tread," man has now been foolish enough to rush in where angels used to fly.

BRUCE MAZLISH

List of Contributors

BRANDFON, ROBERT L., Assistant Professor of History, Holy Cross College; Lecturer, Sloan School of Management, Massachusetts Institute of Technology (part time)

CHANDLER, ALFRED D., JR., Professor of History, Johns Hopkins University

COCHRAN, THOMAS C., Professor of History, University of Pennsylvania

COOTNER, PAUL H., Associate Professor of Finance, Sloan School of Management, Massachusetts Institute of Technology

FOGEL, ROBERT WILLIAM, Ford Foundation Research Professor in Economics, University of Chicago

HUGHES, THOMAS PARKE, Associate Professor of the History of Technology, Massachusetts Institute of Technology

MARX, LEO, Professor of English and American Studies, Amherst College

MAZLISH, BRUCE, Professor of History, Massachusetts Institute of Technology

SALSBURY, STEPHEN, Assistant Professor of History, University of Delaware

Contents

	Foreword Earl P. Stevenson	v
	Preface Bruce Mazlish	vii
I	Historical Analogy: The Railroad and the Space Program and Their Impact on Society Bruce Mazlish	1
II	A Technological Frontier: The Railway Thomas Parke Hughes	53
III	Railroads as an Analogy to the Space Effort: Some Economic Aspects Robert William Fogel	74
IV	The Economic Impact of the Railroad Innovation Paul H. Cootner	107
V	The Railroads: Innovators in Modern Business Administration Alfred D. Chandler, Jr., and Stephen Salsbury	127
VI	The Social Impact of the Railroad Thomas C. Cochran	163
VII	Political Impact: A Case Study of a Railroad Monopoly in Mississippi Robert L. Brandfon	182
VIII	The Impact of the Railroad on the American Imagination, as a Possible Comparison for the Space Impact Leo Marx	202
	Index	217

The Railroad and the Space Program
An Exploration in Historical Analogy

1

Historical Analogy: The Railroad and the Space Program and Their Impact on Society

Bruce Mazlish

The colorful way of stating the purpose of this chapter (and the entire project which it seeks to introduce) is to say that we seek to comprehend the "vengeance of history." What this last means is made clear by the comment of California's Representative George P. Miller. Quoting, during a debate on the space program, a statement by Daniel Webster more than a century ago on proposals that the Government help open up the West — "What do we want with this vast, worthless area, this region of savages and wild beasts, of shifting sands and whirlpools of dust?" — Mr. Miller concludes that "So, whoever says going to the moon and Mars is a waste of money is risking the vengeance of history."[1] A more prosaic way of stating the purpose of this project is to say that it involves an attempt, by historical analogy, to understand and, if possible, to anticipate the impact of the space program on society. More precisely, we shall investigate the impact of the railroad on nineteenth-century society as a "device" whereby we may anticipate the possible and probable impact of the space program on present-day society.[2]

A major difficulty, however, is that historical analogy, while widely used, is little understood. Therefore, the first task will be to examine what is meant by historical analogy; and we must even be prepared to reject historical analogy as a device of anticipation completely. Indeed, there may be no way effectively of avoiding the

[1] Quoted in the *Wall Street Journal,* June 27, 1963, p. 10. Interestingly enough, however, the great Daniel Webster used his oratorical abilities in favor of the Northern Railroad being built in his home state of New Hampshire.

[2] Cf., *Space Efforts and Society. A Statement of Mission and Work,* a document of the Committee on Space Efforts and Society of the American Academy of Arts and Sciences (Boston, January 1963), pp. 28–29.

vengeance of history. In any case, after a preliminary consideration of the concept of historical analogy, I shall proceed to the second part of my task: a summary of the specific investigations into the railroad's impact on society undertaken by the various scholars whose writings follow this introductory chapter. Here, too, it will be well to indicate the gaps and omissions in our coverage. Third, I shall seek explicit analogies with the space effort, and come to some general conclusions as to the shape of further and future inquiries into this area. Throughout, the tentative and exploratory nature of this entire project should be borne in mind.

I. Historical Analogy: An Inquiry into the Concept Itself

It is extraordinary how often historical analogies are used, and how little reflection is given to their usage. As I have remarked elsewhere, there appears to be no theoretical literature on the subject.[3] Analogy seems to be accepted as a natural way of thinking,

[3] See "A Preliminary Consideration of Historical Analogy: Study of the Impact on Society of the Space Effort" (June 28, 1963), p. 1. As I point out there, the bibliography of works in the philosophy of history, covering the period 1845–1957, published by *History and Theory* (Beiheft I), while listing hundreds and hundreds of articles and books, has nothing on the subject of historical analogy. One of the few who has given prolonged attention to this subject is Oswald Spengler. In the *Decline of the West*, Spengler sets up autonomous cultures on the model of what he calls biographic archetypes. Having isolated his cultures as if they were in fact separate organisms, he then goes on to establish what he considers to be a morphology for comparative analysis of his cultures. Spengler makes the interesting distinction between analogy and homology. The first, as is to be expected, concerns functional equivalents, while the latter relates to structural equivalents. Thus, for example, in biology, the lungs of terrestrial and the swim bladders of aquatic animals are homologous, while lungs and gills are analogous. Extended to history, this method, according to Spengler, allows us to go beyond shallow analogies and comparisons, such as that of Napoleon and Caesar, and to perceive, for example, that the homologous form of Napoleon is Alexander the Great. Both of these men stand in the same phase of a declining culture, while Caesar comes later. Thus, for Spengler, Napoleon and Alexander are homologous, whereas Napoleon and Caesar are analogous. While Spengler's theory is suggestive in pointing out that functional and structural likenesses need to be distinguished, it is actually of little assistance to the average historian, who does not accept Spengler's notion of cultures as organisms existing in autonomy and isolation and without real historical change. There are practicing historians, however, like Stanley Elkins, *Slavery: A Problem in American Institutional and Intellectual Life* (Chicago: University of Chicago Press, 1962), who, while not offering a theoretical treatment, do use analogy: in this case the analogy of the Nazi concentration camps to the southern treatment of Negroes. So, too, Eric McKitrick, *Andrew Johnson and Reconstruction* (Chicago: University of Chicago Press, 1960), uses an analogy between the surrender of Germany and Japan in 1945 and the surrender of the South to the North in 1865; and Stanley Elkins and Eric McKitrick, "A Meaning for Turner's Frontier," *Political Science Quarterly*, LXIX (1954) draw an

requiring little reflection. It is simply part of our historical consciousness, whereby, willy-nilly, we "see" present episodes in terms of past experiences. To cite some random examples, in an excellent article on the Russian intelligentsia, the historian Richard Pipes finds it natural to view the diminishing role of the nineteenth-century Russian cultural intelligentsia as analogous to the fading importance of the seventeenth-century boyars.[4] In a different context, discussing the shift in world view from a cosmos under the aspect of hierarchy to one under equality, the historian J. H. Hexter remarks, "The more acute men of the seventeenth century believed that the insights of science and the arguments of Cartesian philosophy alike shattered that world view. With the passion for oneness and *appetite for analogy* (my italics) that has been a persistent trait of Western thought, men were soon shaping their ideas of human affairs after a cosmic model that knew no hierarchy."[5]

More recently, concerning the space program, it is not only congressmen but scholars, social scientists, and administrators who relax naturally into the use of analogies. Thus, two legalists, Philip C. Jessup and Howard J. Taubenfeld, write on *Controls for Outer Space and the Antarctic Analogy,* and Joseph M. Goldsen, head of RAND's social science department, at one point declares that "Just as the discovery of the New World hundreds of years ago profoundly modified the whole course of historical and human orientation of Western Europe for centuries thereafter, so may our present world outlook undergo deeply pervasive change as a consequence of space exploration." Karl Deutsch, a political scientist, reminds us that "surprise attacks confined to targets in outer space might therefore yield limited advantages, somewhat as in seventeenth- and eighteenth-century warfare, when both France and England were incapable of delivering 'first-strike' knockout blows against each other, but were capable of surprise attacks against each other's installations overseas."[6] Similarly, James E. Webb, Chief Administrator, NASA, uses the loose or implied form of analogy when he re-

---

analogy between the experience of individuals in housing projects erected hastily under wartime conditions and the experience of Americans at the frontier. (I owe these last examples to Professor Richard Buel, Wesleyan University.) As already stated, however, such use does not touch on the theoretical aspect of the subject.

[4] *Daedalus,* 92 (Summer, 1960), 492–493.

[5] J. H. Hexter, *Reappraisals in History* (Evanston, Ill.: Northwestern University Press, 1961), p. 116.

[6] *Outer Space in World Politics,* ed. Joseph M. Goldsen (New York: Praeger, 1963), pp. 5, 160.

marks that "the thrust into space will change the ideas and lives of people more drastically than the Industrial Revolution." [7]

In short, all sorts of competent, thoughtful men find analogy a normal, natural manner of thinking. There are, however, dangers attached to "doing what comes naturally." These were pointed out with especial strength by some of the seventeenth-century scientists and philosophers of whom, ironically, Hexter correctly said that they had an "appetite for analogy." Thus, the members of the Royal Society were constantly exhorted to eschew metaphorical and analogical language, and to hold only to plain expressions. Thomas Hobbes, for example, repeatedly complained that analogies were a prime cause of the ambiguity of ordinary language as compared to mathematics. Galileo, as the scholar Leonardo Olschki puts it, believed that mathematics "excludes the insidious fallacies of the common human tongues and the figurative interpretation of phenomena. It helps to avoid the misunderstandings derived from false analogies and makes impossible all the wrong conclusions which depend on analogical inferences and the metaphorical sense of words." [8]

In the eighteenth century, without Galileo's stress on mathematics, David Hume took up the warning flags against facile analogy. As part of his argument concerning cause and effect, he said,

That a stone will fall, that fire will burn, that the earth has solidity, we have observed a thousand and a thousand times; and when any new instance of this nature is presented, we draw without hesitation the accustomed inference. The exact similarity of the cases gives us a perfect assurance of a similar event, and a stronger evidence is never desired nor sought after. But wherever you depart, in the least, from the similarity of the cases, you diminish proportionally the evidence, and may at last bring it to a very weak *analogy*, which is confessedly liable to error and uncertainty. After having experienced the circulation of the blood in human creatures, we make no doubt that it takes place in Titius and Maevius; but from its circulation in frogs and fishes it is only a presumption, though a strong one, from analogy that it takes place in men and other animals. The analogical reasoning is much weaker when we infer the circulation of the sap in vegetables from our experience that the blood circulates in animals; and those who hastily followed that imperfect analogy are found, by more accurate experiments, to have been mistaken.[9]

[7] Quoted in Lincoln P. Bloomfield, ed., *Outer Space: Prospects for Man and Society* (Englewood Cliffs, N.J.: Prentice-Hall, 1962), back cover.
[8] L. Olschki, "Galileo's Philosophy of Science," *The Philosophical Review*, 52 (July 1943), 349–365.
[9] David Hume, *Dialogues Concerning Natural Religion* (New York: Hafner Publishing Co., 1948), p. 18.

And lastly, to take a modern example, the demographer Kingsley Davis points out a typical sort of false analogical reasoning. "The British economist Colin G. Clark," Davis comments, "has contended that rapid population growth stimulates economic progress. This idea acquires plausibility from the association between human increase and industrialization in the past and from the fact that in advanced countries today the birth rate (but not the death rate) tends to fluctuate with business conditions. In today's underdeveloped countries, however, there seems to be little or no visible connection between economics and demography." [10]

Surely, after all these examples, it is clear that historians and laymen alike frequently use historical analogies, and that this usage has its dangers, dangers inherent in all analogical thinking. In order to understand these dangers and defects, however, in relation to historical analogy, we must look a little more closely at the general subject of analogy itself.

Logical Analogy

Analogy is dealt with in all textbooks in logic.[11] As one of them states, "Analogy is the most primitive and, at the same time, one of the most important of all forms of reasoning." Reasoning by analogy seems to lie behind all induction, and can be defined as follows: "Analogy is a process of reasoning whereby we conclude, from the fact that all members of a group are known to have certain characteristics and that some members are known to have other characteristics, that therefore it is likely that the remaining members possess these additional characteristics as well." [12]

Thus, analogy, which originally is based on an "unanalyzed feeling of vague resemblance," is useful in forming hypotheses. To take an example, if certain traits pertain to the earth, Mars, Jupiter, and Saturn as typical planets, we should expect these same traits to be present in the case of Mercury and Venus *even though at first we do not have the actual data* for the latter. Thus, if the first four planets mentioned rotate around an axis, and we know that Mercury and Venus resemble them in revolving around the sun in an elliptic

[10] Kingsley Davis, "Population," *Scientific American*, 209, No. 3 (September 1963), 70.
[11] There seems to be, however, only one full-scale inquiry into the subject, Harald Hoffding, *Der Begriff der Analogie* (Leipzig, 1924).
[12] Frank M. Chapman and Paul Henle, *The Fundamentals of Logic* (New York: Scribner's, 1933), p. 299. Basically, what marks an analogy is the notion of relationship. The latter is primarily either structural, as when the structure of a map corresponds to the structure of the country it represents, or functional, as when we compare gills in a fish and lungs in a man.

orbit, we assume (by analogy) the hypothesis that Mercury and Venus also rotate around an axis. As Cohen and Nagel sum it up, "Analogical reasoning is therefore seen to be a case of probable inference which depends on fair sampling." [13]

The matter of "fair sampling" is important. An analogy does not prove anything. It merely suggests a possibility. If four planets rotate around an axis, that is obviously better ground for assuming all other planets so rotate than if only one planet were originally involved. Of course, the aim of careful inquiry into an analogy is to see whether we can make it so strong that we can generalize from it, that is, can reason "from a group of cases not merely to the next instance but to all other cases." [14]

In such a careful inquiry, we must beware of superficial reasoning. Probability will be meaningful only if we attain some sort of understanding of the "structure" of the situation involved, and keep a keen eye out for possible changes in the situation. Thus, for example, a man who reasoned (without knowing his bank balance) that because four $10 checks he had written were cashed by his bank a fifth $10 check would necessarily be good, would clearly be guilty of fallacious reasoning. Similarly, other limiting factors must be reckoned with in estimating the probability of an analogy.[15]

Another use of analogies, in addition to suggesting hypotheses, is to illustrate and clarify general principles. As one logician puts it, "If you are trying to explain the way a steam engine or a gasoline motor works, it may be helpful to find something similar that works on a similar principle. . . . They [analogies] act as small-scale models of the real thing, and they make an easy first step toward complete understanding." [16]

Thus, to sum up, analogies are a natural and useful way of thinking. They serve as simple models for understanding, and they suggest hypotheses to be tested. In the latter form, they must appeal to probability, that is, the number of cases involved, and they ought to appeal to what we can call "relevance," that is, an analysis of the actual factors involved (such as the bank balance in the case cited earlier). The farthest aim of an analogy is to become a generalization.

[13] Morris R. Cohen and Ernest Nagel, *An Introduction to Logic and Scientific Method* (New York: Harcourt, Brace, 1934), p. 288.
[14] Chapman and Henle, *op. cit.*, p. 303.
[15] For details, see *ibid.*, pp. 300–303.
[16] Monroe C. Beardsley, *Practical Logic* (New York: Prentice-Hall, 1950), p. 106.

Uses of Analogy

Now, from this brief, general discussion of analogies, what conclusions or hints can we draw for our discussion of historical analogy? Without staying too close to the purely logical discussion of analogy in the textbooks cited, we can make the following extrapolations. The first is that analogy is built into the very language constantly used by the historian. Thus, when he talks of "the Space Revolution" (as, for example, Lincoln P. Bloomfield does), or of the Industrial Revolution, he has borrowed his term, by analogy, from the original "revolutions" of the heavenly bodies.[17] Almost always, historical terms of this kind constitute very loose and unexamined analogies; yet they are necessarily the common coin of our historical understanding. One reason for this is that analogies or metaphors establish a relationship, often emotional as well as logical, between otherwise disparate items. Thus, the nineteenth-century terms of "iron horse" and "horse power" serve as transitional symbols, linking animal and mechanical images, and allowing man to bridge over the otherwise alarming gaps in his culture. Only later, with his developing science, does man substitute neutral and purer terms like diesel for the "iron horse," and BTU's or watts for measures of power. Even today's key word "rocket" (used in 1829 by George Stephenson for the name of his famous early locomotor), was developed from the resemblance of the rocket's firework shape to that of a bobbin or spool (in Italian, "rocchetta").[18]

The second thing to note about historical analogy is its function as both a myth and a model. Thus, analogy provides an "original," an archetype, offering us the secure feeling of a familiar experience. For "primitive," or "archaic" man, to use Mircea Eliade's terminology, there can be no act acknowledged "which has not been previously posited and lived by someone else." The life of primitive man "is the ceaseless repetition of gestures initiated by others."[19] Now, of course, for modern man the mythical approach to life has atrophied considerably. Yet many of the same emotional — if not logical — needs are probably satisfied through the comforting device of analogy. For example, what has been called the "cult of antiquity" reassured the French revolutionists that they were merely doing

[17] Cf. Sigmund Neumann, "The International Civil War," *World Politics,* I, No. 3 (April 1949), 336.

[18] On the general subject, see Benjamin Ide Wheeler, *Analogy and the Scope of Its Application in Language* (Ithaca, N.Y., 1887).

[19] Mircea Eliade, *Cosmos and History* (New York: Harper Torchbooks, 1959), p. 5 and *passim*.

what the virtuous republicans of Rome had done before them. Even the dialectic and forward-minded Marxists needed the support of the past, and modeled their portending revolution on the events of 1789 (with a few corrections). As a modern commentator puts it, "In the 19th century, revolutionaries everywhere saw analogies to what had been happening in France between 1789 and 1799. . . . Every upheaval, however insignificant, was interpreted in terms of recent French history, usually misunderstood."[20]

Often the analogical connection is made merely through the use of language itself. To cite only one instance, talking of space as "a new ocean" on which "we must sail" (President Kennedy's words) invokes implicitly an analogy with the fifteenth-century Age of Discovery, as well as arousing all our unconscious associations with that event.[21]

Basic to what I have called the mythical function of historical analogy is the underlying assumption that the past both validates and predicts the future. Thus, it is generally assumed that because something happened in the past (or is asserted to have happened), something analogical in the future should and will also occur. Because America disregarded Mr. Webster's advice and plunged into "this vast, worthless area" with supposed great success, it is assumed that similar success will crown its disregard of the present-day Cassandras who view outer space as a "vast, worthless area." Because the Age of Discovery rewarded some of the participant nations (the later decline of Spain, a prime instigator and initial beneficiary, is often conveniently overlooked), it is assumed that the Age of Space Exploration will have the same beneficient consequences for its participants; and so on. Let me conclude, however, with an interesting variant of this aspect of historical analogy. It emerges from a confusion of the notions of probability and possibility. In the nineteenth century, those who opposed the railroads asserted with great concern and seriousness that speeds as high as sixty miles an hour would addle people's brains. Because this turned out to be wrong (although it was possible), it is now often asserted that, therefore, any speed (and thus, by implication, anything) is possible. The latter assertion about speed, of course, may be true (up to the limit of the speed of light, or just below), but the asser-

---

[20] George Lichtheim, "Reflections on Trotsky," *Commentary*, 37, No. 1 (January 1964), 52. Cf. also the rest of his comment.
[21] Cf. Philip C. Jessup and Howard J. Taubenfeld, *Controls for Outer Space, and the Antarctic Analogy* (New York: Columbia University Press, 1959), p. 210, for the political implications of this analogy.

tion derives no additional support from the analogy to the nineteenth-century "possibility" which turned out to be only a poor probability.

At no point, however, must we underestimate the importance of historical analogy in its mythical function. Analogical myths give not only needed emotional continuity and support, but they pass readily into models. As models, we must notice two points about analogy. The first is what we may call its self-fulfilling aspect. If we assume that the future will resemble the past, and act forcefully on this assumption, there is a good chance that we shall indeed create a correspondence. This is, of course, the familiar position of pragmatism, only stretched now to cover the use of analogy. The second point to notice is that analogical models may, in fact, offer not only inspiration but also precise ways of proceeding. For example, the suggestion has been made that the "U.S. Forest Service's handling of secondary consequences may furnish a useful analogy of a course that NASA may be forced to follow." [22] Obviously, if we were to take the Forest Service's handling of certain secondary consequences as a model for dealing with similar NASA problems, we should then overtly be operating on an analogical basis.

This last, of course, brings us close to the heart of our own enterprise: the study of analogy as a "device of anticipation" or as a guide to "scientific" understanding. Before, however, entering on this last effort, let me briefly summarize what we have said up to now. First, analogy is basic to man's use of language. Second, it is a natural way of reasoning, and basic to inferential thought. In addition, analogy, especially historical analogy, has an important function as myth and as model (the two being at opposite ends of a continuous spectrum). It provides a necessary emotional security and meaning to man's life, as well as having a certain propaganda value based on the former attribute. All these values of historical analogy are neither to be deprecated nor underestimated. Nevertheless, when used carelessly, historical analogy can be a misleading guide. Worse, by establishing a facile resemblance, it may serve to prevent a more critical and analytical approach. It is this last challenge which we must now consider.

[22] Richard Kluckhohn, "Ecological and Bio-Geographical Consequences of Massive Technological Enterprises: The Space Program as a Case Study," mimeographed, February 1963, p. 6. Kluckhohn's additional comment that "past history has shown that secondary forest effects often become primary aims" is most interesting.

Problems in Historical Analogy

Historical analogy, like all analogy, as we have seen, must start as an "unanalyzed feeling of vague resemblance." The purpose, therefore, of close critical and analytical treatment of a historical analogy is to extract, if possible, generalizations that, in the form of hypotheses, can be applied to other cases. Unlike analogy in general, however, historical analogy appears to be faced with two seemingly special problems.

The first emerges from the fact that analogical reasoning is a "case of probable inference that depends on fair sampling." An analogy to only one instance clearly lacks the strength of an analogy to many instances. Yet, this "one instance" quality seems to characterize almost all historical analogies. It is the same problem that is at the heart of the debate about "unique events" in history, and, indeed, of the ubiquitous discussion among philosophers and historians as to the status and nature of history as a scientific discipline.[23] Now, there are ways of bypassing this problem. An analogy from the impact of the discovery of microscopic life, in the seventeenth century, to the impact of the possible discovery of extraterrestrial life may have only one instance; but, obviously, to the general question of the impact of *discovery* as a social phenomenon, there may be many analogies. From these many, we may, in theory at least, derive a generalization that will be useful in analyzing the impact not only of the discovery of microscopic and extraterrestrial life but also of the discovery of atomic energy, the huge nature of space, and so forth. Thus, without underestimating the difficulties in obtaining "fair samples" in history, we need not despair completely.

The second problem possibly unique to historical analogy is the unusual allowance which must be made for what can be called "historically conditioned awareness." For example, once the railroad has shocked people into a new awareness of speed — vehicles traveling at sixty miles an hour — the impact upon their imaginations and sensibilities of space speeds of six thousand miles an hour will be vastly diminished. Hence, superficially analogous impacts may differ sharply in terms of one major variable: which event came first and thereby conditioned man's historical awareness. As one historian puts it (using as his example men's reactions to the Industrial Revolution as conditioned by their prior passion for the goals of happiness

---

[23] For example, see Carey B. Joynt and Nicholas Rescher, "The Problem of Uniqueness in History," *History and Theory*, I, No. II (1961).

and equality preached by the French Revolution), "Events, facts, data, happenings assume their significance from the way in which they are experienced. To be kicked about was for a slave in antiquity or for a Chinese coolie only yesterday a matter of course, but the birching of a schoolboy in our day is considered by most people in the West as something absolutely horrifying." [24] Obviously, the impact of inflation or unemployment on the United States may be significantly different before and after the New Deal legislation and the new values and awareness that the latter embodied.

With these cautions in mind, I am tempted to state categorically that, *for purposes of scientific knowledge*, only a historical analogy that (1) allows for progressive trends, and (2) rises above the comparison or resemblance of two simple elements can be of any real value. What I have in mind is that we cannot really understand a single item, as to its structural or functional position, without studying the system, both dynamic and static, within which it rests. Therefore, for example, a comparison of the effect of the discovery of microscopic and extraterrestrial life can only be meaningful in terms of a systematic study of the two societies in which the effect is felt.

Resisting, however, the temptation to make such a strong, categorical theoretical claim as the one above, I shall simply make a specific suggestion along these lines concerning our immediate research problem. Thus, in my judgment, in an effort to understand the secondary consequences of the space program, the most significant analogy to its impact would be one that treats the space program not merely as an isolated matter, say, of exploration, or military preparedness, or scientific innovation, but as a complex *social invention*. By social invention, I mean an invention that is technological (e.g., missiles, launching pads), economic (e.g., involving large-scale employment of manpower, widespread use of materials), political (e.g., involving new forms of legislation, and new dispositions of political forces), sociological (e.g., affecting kinship groups, communities, classes), intellectual (e.g., changing man's views of space and time), and so forth. My definition is admittedly broad; but I believe it is the only one that does justice to the reality of the space program and its impact on society.

The specific analogy I have suggested, and which is exemplified by the chapters that follow, is the coming of the railroads. This is a social invention, commensurate with the space effort in its wide-

[24] J. H. Talmon, "The Age of Revolution," *Encounter, XXI*, No. 3 (September 1963), 14.

ranging effects: the nineteenth century has appropriately been called the Railroad Age, as the twentieth is now being called the Space Age. Further, as I have remarked in my earlier paper, the railroad is sufficiently part of the Industrial Revolution to involve us with an industrial society (whereas the boundary conditions for, say, the Age of Discovery would have been vastly different) and yet sufficiently far in the past to allow us a decent perspective; it also provides us with a great deal of solid historical data.

Obviously, no two social inventions can be absolutely comparable in their impact on society; however, the differences can be as illuminating as the similarities. In any case, a study of the railroads, as analogous to the space program, may provide us with generalizations useful in projecting research into the space program's future impact, as well as of the impact of all other social inventions, past and present. In short, we are really attempting to set up a new branch of comparative history: the study of comparative or analogous social inventions and their impact on society.

Now, the first thing that must be said of this effort is that it is subject to all of the difficulties involved in comparative history in general. To go into details on this latter subject here, however, would be unwarranted as well as excessively difficult. The magnitude of the task will be clear to anyone familiar, for example, with the problems involved in the study of comparative feudalisms.[25]

Rather than entering on a discussion of comparative history, per se, I shall restrict myself to some observations on the specific topic of social invention as a form of historical analogy. There are three points I shall try to make. The first is that no satisfactory theory or even typology on the subject seems to exist. There is a fine pioneering work, *The Sociology of Invention,* by S. C. Gilfillan, dating back to 1935, but its emphasis is on the social *causes* of *technical* invention (although it does have a few pages on social results). Nevertheless, two of its generalizations seem especially worth pondering over and testing further:

[25] Cf. Bert F. Hoselitz, "On Comparative History," *World Politics,* IX, No. 2 (January 1957). What Hoselitz has to say about comparative history should be compared to some of my comments on historical analogy: "The application of the comparative method to historical material is not possible if it is confined to a comparison of chains of events, since these are essentially unique in character. . . . In order to apply the method of comparison to historical material it is necessary that social systems (or parts of social systems) be compared with one another, or that developmental sequences, expressed in terms of generalizable variables, be the object of comparative study." Hoselitz, *op. cit.,* p. 274. In general, see also the periodical, *Comparative Studies in Society and History.*

1) *Equivalent invention:* perceived needs are met by various *unlike*, as well as duplicate solutions, so that any great invention is simultaneously paralleled by other, often utterly dissimilar means for reaching the same end at the same time, e.g., reaching California by clipper, steamer, pony express, railroad and telegraph. Inventions may be seen as arriving in functional groups.

2) Hence *no* single *invention* ever *revolutionizes civilization,* nor brings, simply thru having been invented, any important changes in the life of the mass of men.[26]

The whole of Gilfillan's book is exceptionally worth reading, bearing in mind the limitations of date and focus that have been noted.

Other, partial treatments of our subject can be found in books on social change. Classics in the field, such as William F. Ogburn's *Social Change* (1922), or Robert M. MacIver's *Society, its Structure and Changes* (1931), are suggestive and certainly worth reading. However, like their more modern successors, they only nibble around the edges of our particular problem.[27] The impact of a social invention, as we have defined it, and certainly the question of historical analogies on the subject, is not given significant attention, and the inquiry is left formless.

What shape a typology or theory (or both) of social invention and its impact on society should take is, as yet, difficult even to hint at. On one level, one might specify a statistical comparison as a first basis for analogy. For example, the railroad and the space impact might be compared in terms of such items as (1) the absolute number of workers employed as well as the relative percentage of the manpower available in each society that they constitute; (2) the number of new communities and their sizes; (3) the percentage of GNP investment; and so forth.[28] While essential, however, this sort of statistical analogy is only a small part of the picture, and must be interpreted against a larger, qualitative context. It is the latter that provides the greatest theoretical difficulties. For

---

[26] S. C. Gilfillan, *The Sociology of Invention* (Chicago: Follett Publishing Co., 1935), p. 12.

[27] Wilbert Ellis Moore's book *Social Change* (Englewood Cliffs, N.J.: Prentice-Hall, 1963), is a typical modern example, disappointing, however, in what it has to offer relevant to our subject. See, however, Lewis Mumford, *Technics and Civilization* (New York: Harcourt, Brace, 1934), another pioneering effort in English. The recent grant of $5,000,000 by IBM to Harvard, to study the social consequences of technological change, especially of automation, illustrates the increasing attention being paid to this as yet relatively uncultivated field.

[28] Some of these figures, in fact, are given in the several chapters that follow; see especially Cootner, Fogel, and Cochran.

example, in order to understand the consequent impact, we shall probably want to ask (while using statistics for the answer wherever, and as far as, we can), which groups in the society welcomed and worked for the social invention and which ones opposed it. Why did these groups take up their particular stance (the economic, political, social reasons, etc.)? What, then, happened to the expectations of these various groups, and how did their various stances affect the impact of the social invention on society (obviously, if primarily military men support and activate the space program, the effect on society may well differ from one where civilians provide the primary direction)? Clearly, even in the simple instances mentioned, the complexity of the problem is great. Sophistication and subtlety are needed to set up the conceptual apparatus. For example, it is obvious that one will want to consider the technological and economic impacts on society that did in fact occur; it may not be so obvious that one will also want to weigh the technological and economic impacts that did *not* occur (for example, the development of the railroad obviated the wider development of the canal system; cf. Cootner, Chapter 4 of this volume).[29]

Forgoing here the audacious attempt to set up, a priori, a typology or theory of the impact of social inventions — this ought only to emerge from empirical work, such as in the chapters following, and from similar historical-analogy studies, that is, increased "fair sampling" — I pass on to my second point. This is simply that the study of social impact may be otherwise phrased as the study mainly of secondary consequences (and tertiary consequences, and beyond). The immediate impact of the railroad or the space program is probably obvious: the railroad in the 1840's used X tons of iron rails; the space program in 1965 employed X number of technicians and scientists. But the secondary effect, in both these cases, on the overall economy or on the nation's educational system, is not clear. It is these secondary consequences, or ramifications of the primary impact, that may have the greatest significance for society, and

[29] Something of this last is involved in the debates over the value of war, as fostering invention and development, in spite of the high price in life and morality. Thus, the contention is often made that war stimulates certain developments that would not otherwise occur, or would not occur so rapidly, and this contention is made without weighing the negative impact. In so far as the space program may be considered as part of defense expenditures, this debate may be especially pertinent; I say this, however, without passing any judgment on this complex issue. See J. Nef, *War and Human Progress: An Essay on the Rise of Industrial Civilization* (Cambridge: Harvard University Press, 1950), for a scholarly treatment of the question as it relates to war.

that need the most concentrated study. The hope, of course, is that once we understand more about these consequences — initially "unintended consequences" — for example, through the derivation of useful generalizations from historical analogies, the more we shall be able to do to control them.

Ironically, foresight and control may be a two-edged weapon. It is obvious that many desirable consequences (as judged by large parts of the population) are achieved through ignorance, and, if known ahead of time, might well have been thwarted. Adam Smith was one of the first to point out the importance of "unintended consequences." Explaining the transition from feudalism to capitalism, he concluded that

A revolution of the greatest importance to the public happiness was in this manner brought about by two different orders of people who had not the least intention to serve the public. To gratify the most childish vanity was the sole motive of the great proprietors. The merchants and artificers, much less ridiculous, acted merely from a view to their own interest, and in pursuit of their own peddler principle of turning a penny wherever a penny was to be got. Neither of them had either knowledge or foresight of that great revolution which the folly of the one and the industry of the other, was gradually bringing about.[30]

With foreknowledge, would the "great proprietors" have allowed the "revolution" to have occurred? It is difficult to know. Familiar to us in terms of Hegel's "cunning of reason" and Marx's "dialectical materialism," the unintended consequence can only function properly when it is truly unintended. If, to take a contemporary example, the space program were predictably to have certain secondary consequences related to race relations, to labor conditions, or to the disposition of political power, would the "losing" groups allow the development to proceed to the critical point? The question is an involved one. In any case, humanity seems generally convinced that foresight and control over social processes are useful gifts. Whether in particular circumstances we choose to exercise those gifts, and in what ways, is a matter of our deepest values. And a discussion of the latter is beyond the scope of this introduction.

The third, and last, point I wish to make here concerns the complexity of our problem, and the consequences this has for the method we ought to employ in our researches. One difficulty in the social

[30] Adam Smith, *The Wealth of Nations* (2 vols., London: Everyman's Library, J. M. Dent & Sons Ltd., 1954), Vol. I, pp. 369–370 (Book III).

or human sciences is the discovery that the more we come to know, the more we see how interrelated and connected, in more and more complicated fashion, are our data.[31] Perhaps a familiar point by now, this observation bears repetition and stress. The phenomena of social and historical existence ramify in an almost bewildering fashion. To take an instance of what I mean: in his chapter on "The Social Impact of the Railroad," Thomas Cochran examines the effect of the railroad on family structure. His assessment is complicated by the fact that, on one hand, the railroad strengthened family ties because of increased possibilities of communication, while, on the other hand, it weakened family ties by extending individual mobility. How are we to assess the over-all impact of the railroad on family structure? Actually, up to this point, the problem has been posed in relatively simple terms. As some of the other chapters show, the railroad had other secondary consequences: for example, it affected farm life and farm economics, it brought about new economic arrangements favoring urban concentrations, and it fostered an industrialized society with new values and new conceptions about romantic love and family life. How did these consequences of the railroad strengthen or weaken family ties? How do we weigh their relative effects? Worse, how do we trace the railroad's second-order effects on, say, economics in general, the effect of these general economic conditions on social life in general, and then the effect of these general social changes on family ties in particular?[32]

Obviously, any simple method of tracing unitary cause and effect will not do. Ideally, we need some sort of multivariate scheme of causation. What I have in mind can be illustrated by Seymour Lipset's remarks on the relationship of economic development and democracy. Thus, he says in his book, *Political Man*:

> It would be difficult to identify any *one* factor crucially associated with, or 'causing,' any complex social characteristic. Rather, all such characteristics (and this is a methodological assumption to guide research, and not a substantive point) are considered to have multi-variate causation, and consequences. The point may be clarified by a diagram of some of the possible connections between democracy, the initial conditions associated with its emergence, and the consequences of an existent democratic system.

---

[31] See the discussion in the magazine of *The American Association of University Women* (May 1964).

[32] Modern computers may aid the social scientist in his effort to assess and weigh the interrelated effects. However, it is not clear as yet exactly how much aid will really be forthcoming.

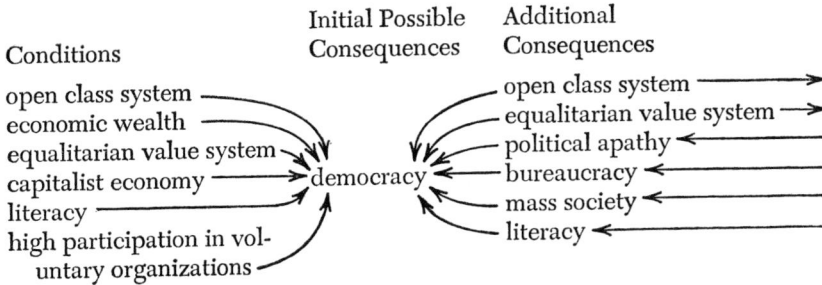

The appearance of a factor on both sides of 'democracy' implies that it is both an initial condition of democracy, and that democracy, once established, sustains that characteristic of the society — an open class system, for example. On the other hand, some of the initial consequences of democracy, such as bureaucracy, may have the effect of *undermining* democracy, as the reversing arrows indicate. Appearance of a factor to the right of democracy does not mean that democracy 'causes' its appearance, but merely that democracy is an initial condition which favors its development. Similarly, the hypothesis that bureaucracy is one of the consequences of democracy does not imply that democracy is the sole cause, but rather that a democratic system has the effect of encouraging the development of a certain type of bureaucracy under other conditions which have to be stated if bureaucracy is the focus of the research problem. This diagram is not intended as a complete model of the general social conditions associated with the emergence of democracy, but as a way of clarifying the methodological point concerning the multi-variate character of relationships in a total social system.

Thus, in a multi-variate system, the focus may be upon any element, and its conditions and consequences may be stated without the implication that we have arrived at a complete theory of the necessary and sufficient conditions of its emergence.[33]

Equally obvious is the fact that this ideal method will be extremely difficult to honor in practice. Yet I have not pointed out the lack of a general theory of the impact of social invention, the complexity of the materials, and the problems of methodology in order to dissuade us from a study of the pertinent historical analogy. The task is too important for our society to neglect. Rather, I have sought simply to underline the nature of the problem confronting us so that we can proceed most effectively. Historical analogy as a

[33] Seymour Martin Lipset, *Political Man* (Garden City, N.Y.: Doubleday, 1963), pp. 61–62. Cf. Paul F. Lazarsfeld, "Interpretation of Statistical Relations as a Research Operation," in P. F. Lazarsfeld and M. Rosenberg, eds., *The Language of Social Research* (Glencoe, Ill.: The Free Press, 1955) pp. 115–125; and H. Hyman, *Survey Design and Analysis* (Glencoe, Ill.: The Free Press, 1955), Chapters 6 and 7.

linguistic device, and as myth and model, will always be with us. Historical analogy as a scientific device, emerging from the vague intuitions of myth and model, is, potentially, a most valuable tool. The caution is that we must accept only detailed, informed studies of the complex phenomenon itself, and not accept simple or fuzzy generalities about vague resemblances. Empirical detail, sometimes boring perhaps at first glance (and perhaps last glance, too, alas), sometimes far removed from our present anxieties, is the only basis on which useful generalizations, instead of generalities, can be formulated. It is, ultimately, the interplay of theory and of empirical detail, each inquiry arising progressively from the other, that can give us the means of understanding, and thereby perhaps controlling, important areas of our social existence. What follows concerning the railroad and the space impact is a most tentative step in that direction.

II. The Railroad as a Social Invention Analogous to the Space Program

In discussing historical analogies, I have suggested three criteria, among others: (1) they should be based on detailed, informed empirical studies; (2) they should preferably be concerned with the complex relationships obtaining in a large system, rather than with the simple comparison of two isolated elements; and (3) they should appeal to as large a "fair sampling" as possible.

The first criterion means that there is no substitute for the chapters on specific topics that follow. While I shall try to give brief summaries of them, they themselves speak in a unique voice that I cannot imitate or encapsulate. What I can attempt here therefore is mainly to highlight themes, to point out interconnections, and to suggest extrapolations. Moreover, there is a real question whether the reader is not best advised to return to this part, as a form of conclusion rather than as an introduction, after having read the specific chapters.

The second criterion implies that the best historical analogy is one which attempts to deal with a "total" system. In practice almost all studies, and certainly the present one, are limited. Thus, in studying the railroad's impact on society, our attention has been focused mainly on technological, economic, managerial, social, political, and intellectual aspects, and even on these, of course, in more or less differing degrees. Other, neglected aspects which come quickly to mind, merely as examples, are the railroad's impact on military affairs; and on such diverse phenomena as art, architecture, and artistic taste. Aside from indicating some of the gaps, however, I shall only be able to add a few remarks on these matters here and there.

The third criterion, closely connected with the previous one, reminds us that other comparisons would be valuable. On the highest level, of course, these would be comparisons with other social inventions — the automobile, or atomic energy — and their impact on society. On our own level, however, of the railroad itself, we would find studies of the experiences of countries other than America — the focus of our work — most useful. For example, Belgium, unlike America, built its railroads as state enterprises, somewhat akin to our space program. And, as Thomas Hughes points out, the American railroad system, in contrast to the British, had as one aim the penetration of a vast interior with unknown and untapped resources, an unexpected consequence being a difference in the engineering techniques applied — the American resorted to temporary, expedient engineering.[34] Thus, instances other than the American railroad would give us a better "fair sampling" in our effort to understand the factors involved in a social invention's impact on society.

With these limitations in mind, I want now to turn to the impressive empirical studies that follow. At the end of a brief overview (meanwhile refraining from overt analogies to the space program), I shall try to set forth some generalizations that might apply to all social inventions, and then turn finally to some remarks on the space effort itself as a social invention.

1. *The Technological (Hughes)*

The word "technology," as used in the late 1820's, began to have attached to it the notion that it was an agent of change.[35] At about the same time, actually on September 15, 1830, the Liverpool and Manchester Railway was formally opened, and "The Railway, a public carrier of passengers and freight on rail by mechanical traction under statutory authority, stood plain for all to see." [36]

Naturally, as Hughes and others point out, the railroad did not,

[34] Hughes, pp. 56–57 of this book.
[35] It is interesting to note that the term "industrial revolution" seems to have been invented by a French economist, Jérome Adolphe Blanqui, in 1827.
[36] Michael Robbins, *The Railway Age* (London: Routledge and Kegan Paul, 1962), p. 14. Robbins' book, along with Cyril Bruyn Andrews, *The Railway Age* (New York: Macmillan, 1938), is the most useful work on the British railroads' impact on society. For America, a comparable work is Seymour Dunbar, *A History of Travel in America* (4 vols., Indianapolis, 1915), see especially Volume III.

any more than any other technological invention, emerge "full-blown, like Minerva from Jove's forehead." The railroad was a *system*, not a single invention, and it involved a *complex* of elements.[37] Starting as a means of transporting coal from the mine head on rails (often, at first, wooden), then adding the idea of a steam locomotor, then carrying over building techniques from canal technology, and so on and on, the railroad developed gradually and complexly. Forging the usual, and misleading, attention to the motive power element (the steam engine), Hughes boldly and imaginatively focuses on the development of the railroad as a conquest by man of a natural frontier and the consequent creation of a more man-made environment.

Briefly, Hughes' thesis is that the *railway* marks man's conquest of natural obstacles, such as freezing conditions (e.g., in waterways) and difficult elevations (e.g., the problem of gravity and friction), resulting in a steady communication system unaffected by nature's variety. As he so well puts it, "land surfaces that undulated, weather conditions that frustrated, distances that consumed inordinate time, must all give way to a fixed, uniform environment conducive to an economic system involving co-ordination and prediction." Once achieved, the knowledge of how to create this man-made environment passed into the new technological institutions of education — and Hughes supplies details of this impact — and then served as inspiration for the engineers who sought to solve the new problem of intra-urban transportation (brought on partly by the high density of workers) by cutting tunnels and erecting elevated structures in the image of the earlier railroads.[38]

While this is Hughes' main thesis concerning the impact of the railroad on society, there are a wealth of other suggestions in his report. For example, he is aware that the railroad is a *transportation system*, and as such had to be economically competitive with the canals and turnpikes, and that this economic fact affected railroad technology.[39] Also, as a more assured transportation system the railroad had the economic consequence of providing certainty of supply and thus diminishing fluctuations in prices. He reminds us that the railroad brought in its wake widespread technological diffusion: for example, its bridge-building techniques, as with James B. Eads'

[37] For details of this, relating to the British experience, see C. B. Andrews, *op. cit.*, Chapter 1.

[38] It is possible, however, as Dean Gordon Brown of M.I.T. suggests, that the railroad engineers' penetration of the technical schools may also have had adverse results both for railroad technology and for the educational institutes.

[39] Hughes does not leave this, however, as a mere generality. For example, for details see p. 54 and his footnote 10.

construction across the Mississippi at St. Louis, and the pneumatic drill frame developed for the Mt. Cenis tunnel.[40] And he specifically reminds us that the very problem of urban congestion and intra-urban transportation needs was partly brought about by the growth of the railroads and their role in transporting rural dwellers to areas of high population density; later, in the chapter by Cochran we shall see this spelled out in more detail.[41] Hughes is even open-minded and flexible enough to modify his thesis concerning man's construction of a self-made environment by conceding that, "because of the state of technology and for economic and political reasons" peculiar to America, a compromise with nature was effected.

In addition, there are a few points that I should like to extrapolate. One is that, as Hughes points out, the crossing of the Allegheny Mountains by railroad could only come about when the engineer had perfected his conquest of natural elevations by the use of such elements as an "inclined plane" in conjunction with various power sources; it bears noting that this technological development was necessary for the large-sized railroads that then required, as Chandler and Salsbury will show us, new administrative structures.[42] Another is that the conquest of a technological frontier, and the creation of a man-made environment, carries with it large-scale though perhaps vague consequences for man's view of nature; for example, instead of nature threatening to obliterate us, we may destroy it. Lastly, and connected with the changing view of nature, there is the philosophical implication for man as to the nature of the universe and his role in it. I shall revert to this subject when I discuss the views of Hannah Arendt on man's conquest of space.[43]

2. *The Economic* (*Cootner and Fogel*)

The major thesis in this area is that, when viewed in terms of refined and sophisticated economic theory rather than popular imagination, the economic impact of the railroad in America, while important, was not *as* important as generally believed. The railroad

[40] Hughes, pp. 62, 63. Brandfon, on p. 185, mentions Eads' work on the port of New Orleans. One wonders what technological carry-overs Eads brought with him on these two tasks.

[41] Hughes, pp. 66, 67; Cochran, pp. 163–168.

[42] Hughes, p. 59; Chandler and Salsbury, *passim*.

[43] Further work on railroad technology itself might give additional attention to such matters as the steam engine; the development of the electric telegraph — as a device both for safety and for the rapid supplying of data; the introduction of air brakes; and even to such seemingly peripheral developments as the invention of standardized, printed tickets. More remotely, one might choose to study such secondary technological effects as the change in methods of producing steel, if any, brought about by the railroad's demand for a certain quality of steel, and so forth.

was not "the exogenous force which stimulated the pace of economic development," but rather followed the pattern of general investment. The pace and magnitude of the railroad's impact can only be viewed correctly in terms of its being simply an alternate means of transportation, going through stages of adoption, and subject to competition from other means of accomplishing similar ends. For example, it is well to remember that as late as 1840, more miles of canals were still being built in the United States than of railroads.[44]

Cootner and Fogel do not leave the problem of "pace and magnitude" as a generalization; they supply the concrete detail of cycles of construction and percentage of GNP. They remind us that while the railroad aided some areas and fostered some developments it blighted others.[45] They state that the maximum economic effect of the railroad came about fifty years after the first railroad was built, and that even at that time railroad investment accounted for only 2.2 per cent of GNP (it is interesting to note that space expenditures in the first few years have been running around this same figure).[46] Even the secondary impact, they point out, does not necessarily fall on domestic industry.

Like Hughes, they remind us that technology and economics are radically intertwined: as technique improved, the railroad's competitive position vis-à-vis canals and turnpikes slowly bettered, and as cost factors cast their shadow on the railroad, new techniques were devised. Thus, while new locomotors might be technically satisfactory, their engines might not prove sufficiently economical, and more development would be necessary. External economies also played their role. The reduction in relative costs of iron and machinery that characterized the entire economy also gave the railroad a competitive advantage, and fostered additional technical developments employing the cheaper materials.

Cootner and Fogel, too, point out that the railroads helped in centering population in the cities; they then connect this population shift with the need for substantial quantities of coal, replacing wood as a fuel, which the railroads, of course, were also in an excellent position to supply. Cause and effect, here as elsewhere, are almost impossible to separate. Other secondary effects to which they call our attention are, for example, the railroad's impact on the quality of steel production (with the subtle remark that "lower quality" is

---

[44] Cootner, p. 109.
[45] Fogel, p. 105; Cootner, pp. 112–116.
[46] For tables concerning space expenditures, see for example *The American Behavioral Scientist*, VI (March 7, 1963), 14–17.

not an invidious term when the result is adequate to the job intended and the costs are lower; see Chapter 4, p. 120), and the railroad's contribution to the growth of a capital market (a theme to be further developed by Chandler and Salsbury).

On the level of pure theory, Cootner and Fogel offer an interesting argument against the wide-ranging thesis of W. W. Rostow concerning a so-called period of "take-off." According to Rostow (in *The Stages of Economic Growth*), a developing nation reaches a point where a few very rapidly expanding industries push the economy to a plateau from whence "self-sustained" growth then proceeds automatically. In the nineteenth century, Rostow maintains, the railroad was the key expanding industry. Fogel, especially, challenges this thesis, both as to the data it uses, and as to its theoretical assumptions. The debate, of course, has important consequences for our views on the underdeveloped countries of today and on the direction in which they may move, as well as for our convictions about the automatic progress of our own "developing" economy.[47]

All in all, the chapters by Cootner and Fogel offer us the sobering argument that the railroad's impact on the economy, and inferentially the impact of any innovation, is not as great or as pivotal as our imaginations would have it. Indirectly, of course, this seems to suggest, to paraphrase Pascal, that the imagination has its reasons of which the economy is unaware.[48]

[47] See Robert William Fogel, *Railroads and American Economic Growth: Essays in Econometric History* (Baltimore, Md.: The Johns Hopkins Press, 1964), Chapter 4, especially pp. 8, 14, 32–33, 37.

[48] Pascal's comment, of course, was that "The heart has its reasons that reason knows not." As for the Cootner and Fogel thesis, it is not accepted by all economic historians — indeed, it arouses fairly violent reactions — and further work in this area ought to come more directly to grips with the opposing arguments. Also, comparative studies of the specifically economic impact of the railroad, say in England or Germany, would be useful, as well as of other social inventions, for example the automobile. In this connection, the interested reader should see such writings as L. H. Jenks' classic article, "Railroads as an Economic Force in American Development," *Journal of Economic History*, IV, No. 1 (May 1944), 2, where he remarks that "Once railway projects have been conceived and plans for their execution elaborated, it becomes easier for other innovating ideas to be entertained . . ."; and the recent article comparing British and American experience by B. R. Mitchell, "The Coming of the Railway and United Kingdom Economic Growth," *Journal of Economic History*, XXIV, No. 3 (September 1964), 322, where, although he qualifies himself later, the author states that "expenditure on railway building was, at its peak, about two-thirds of the value of all domestic exports, and half as much again as the value of cotton goods exported; or, to find yet another yardstick, it was over twice as great as the maximum level of the Bank of England's bullion reserve in the decade. Such a level of expenditure, reached in such a comparatively short time, was bound to have an enormous impact on the economy. . . ."

3. *The Managerial (Chandler and Salsbury)*

The thesis here is that the railroad was instrumental, indeed crucial, in creating modern industrial administration. The railroad had this role largely because of its unique size and complexity. Operating hundreds of miles of line, with, for the time, enormous capital investments (in 1873 the Pennsylvania Railroad, for example, had an investment of $400,000,000), the railroads had to invent totally new administrative devices. The managerial innovations, as Chandler and Salsbury so well point out, came in stages and differed from company to company.

For example, the large distances and complicated traffic arrangements created problems of safety and of scheduling. To solve these problems, a strict administrative hierarchy, run by professional administrators (who, at first, were engineers), was necessary. So, too, the large capital investments, representing heavy fixed costs, made it requisite to discriminate between profitable and unprofitable services, and this made the collection of accounting statistics essential. A nice combination of the two needs, for safety and profitability of operation, was effected, for example, by the Erie Railroad General Superintendent Daniel McCallum's use of the telegraph both as a means to make trains safe and as a device, by its quick reporting of vital data, to improve co-ordination and administration. Here, we have a convincing example of the fusion of technological and administrative innovation.

Chandler and Salsbury then trace the development of a decentralized administration on the Pennsylvania Railroad, with its division of line and staff duties, so admirably suited for the great distances on American railroads. Such a development, however, was not inevitable, as the New York Central's growth as a centralized bureaucracy, more on the English short-distance model, so well demonstrates. In fact, with its creation by financiers and politicians, more than by professional engineer-managers, the New York Central indicated the future direction to be taken by other American railroads such as the Illinois Central (whose political impact in Mississippi will be studied in the chapter by Brandfon).

Centralized or decentralized, managed by financiers or engineers, the American railroads created the modern form of administration, wherein business activity moved away "from organizations run by entrepreneurs with the aid of personal trustees, relatives, and the like to corporations with a systematized bureaucratic management." Further, by contributing forcefully to the creation of a securities market, the railroads further pioneered in the separation of ownership and management which has become so characteristic of the American corporation.[49] This trend was accentuated by another financial device brought about by the railroad development: the holding company, a legal device to allow for greater control and ownership of subsidiary holdings by a parent company. This came about because, as the railroads expanded toward the West, it became more necessary for them to control rates and to make secure the arrangements with feeder lines. These results could be achieved either by interrailroad alliances, favored by the New York Central, or by actual control of the feeder lines, opted for by the Pennsylvania Railroad. The chosen device used by the latter became the holding company, an invention which then paved the way for the rest of American industry. Incidentally, it seems to have been the "accidental" presence of Jay Gould's cutthroat manipulation of the Erie Railroad that, breaking down the New York Central's attempt to deal with the situation administratively by alliances, won the day for the firmer control by the holding company device.

At a slightly further remove, according to Chandler and Salsbury, the new forms of professional management led to a gulf, not only between owners and managers, but between management and labor. With the necessity for administrative expertise, the Horatio Alger rise from roundhouse to corporation president was no longer a meaningful reality. With the acceptance of this new reality by the workers, the way lay open to the creation of powerful railway unions. The existence of the latter, in conflict with professional management, produced as one further consequence the growth of governmental machinery for mediation. As Cootner remarked even earlier, the railroad with its state and federal subsidies, its rate-setting needs, and now with its labor problems, fostered the growth of government regulatory commissions.[50] And so, as a result of the administrative innovations first brought about in the railroads because of their unusual size and complexity, the web of bureaucratic connections unfolded from them to other industries and then to the government itself.

4. *The Social (Cochran)*

Whereas Chandler and Salsbury look at the professional managers as an administrative innovation, Cochran is concerned with them as

---

[49] See Adolph Berle and Gardiner Means, *The Modern Corporation and Private Property* (New York: Commerce Clearing House, 1932).
[50] Cootner, p. 122.

a new social group, possessed of a new status. As a group, moreover, he views them as characterized by a concern for "wise expansion and efficient operation rather than profit per se." [51] Also from the social side, we may note that, for a time at least, the railroads made it possible for executives and administrators to live in the suburbs and commute leisurely to the overcrowded city. Thus, the city's own division into the "right" and the "wrong" side of the tracks was extended to the fashionable suburbs.[52] Cochran also reminds us that the railroad, by pioneering in the stock market, helped to create a group of middle-class security holders, whose "conservative" living habits and political views presumably overlapped with the new class of professional managers. Thus, the market mechanism, which holds Cootner's attention from the economic side, and Chandler and Salsbury's attention from the administrative side, also has important social repercussions.

Demographically, Cochran points out, the railroad's results were complex. On one side, it attracted workers from the country to the city. They came partly because industry was in the city; and industry was in cities like Chicago because converging railroad lines in fierce competition with one another often set low freight rates in these new urban centers. But partly the workers came off the farm because the railroads raised the levels of expectation in relation to material success at the same time that they diminished the self-sufficiency of the rural areas. On the other side, however, the railroads also contributed to the location of population in areas that had not been agriculturally profitable with earlier forms of transportation. Thus, instead of filling in the more easterly states, the railroad created "pockets" of population, often of immigrants encouraged to come to America by railroad agents in Europe, in the states between the Mississippi River and the Rockies. As Cochran remarks, it is an open guess whether this deployment of population was economically or socially desirable. In any case, it is certain that the railroad had a pronounced demographic effect in terms of spacial mobility and redistribution of population.

In addition to institutional and demographic effects, which, of course, need more intensive study and exposition, Cochran's discussion raises two other interesting questions. The first concerns the role of government in railroad building and regulation. Surprisingly

---

[51] After the 1850's, however, as Chandler and Salsbury point out, many of the railroads were run at the top by financiers, who displaced the professional managers.
[52] Cf. Robbins, op. cit., p. 51.

enough perhaps, until 1857, there was a good deal of sentiment in favor of government construction and ownership of the railroad. In any case, some sort of involvement of the railroad with government was unavoidable. At first this was mainly in connection with the states, for the railroads required franchises and the ability to condemn property. They also needed money, and by 1835–1841, for example, Massachusetts had extended $4,000,000 of its credit, or over half the capital invested in the Western Railroad.[53] After 1850, it was the federal government that loomed large, especially for mail contracts and land grants. Vast blocks of public land were handed over to the railroads, until about 1872 when the realization that the railroads were selling the lands at increased prices helped put an end to the policy.[54] Whether state or federal, however, the railroads' relations to government were often crucial, and the result was the growth of professional lobbyists, their practices as much an innovation as the air brake. (Later, we shall see in Brandfon's chapter how important, for good or for bad, their role in state politics might be.) We might also remember that political activity — and I would include here the courts and legal codes as well — is a critical part in the acceptance and development of almost any social invention.

The second question, or group of questions, that Cochran raises has to do with the social-psychological impact. Here he deals with raised levels of expectation (cf. Chapter 7 for the importance of this in Mississippi), with the railroad's impact on religious beliefs, and with the railroad's effect on our conception of time, especially in its increasing the tempo of life, and introducing standard time. Incidentally, one effect of the railroad's "speed-up" of time was economic and administrative: businesses now required much smaller inventories.

Over-all, as Cochran reminds us, the railroad's impact on our social-psychological nature was largely in tune with existing American values. These values, of achievement, efficiency, progress, etc., favored the reception of the railroad, and, in turn, were reinforced by the railroad. As with the technological and economic "frontiers," so in the social-psychological, there was an American "way" or "style" of acting in relation to the coming of the railroad. That there were other components to the American psyche, less positive or favorable to the penetration of America by the railroad, we shall

[53] *A Documentary History of American Economic Policy Since 1789*, ed. William Letwin (Garden City, N.Y.: Doubleday, 1961), p. xviii.
[54] *Ibid.*, pp. xix, xxi; Cochran, p. 173.

see when we discuss the literary and imaginative impact of the "iron horse" as presented by Leo Marx.

Undoubtedly, the social impact of the railroad was enormous — but diffused. What job images did it create for the young, and what work patterns for the old? What effect did the mobility of workmen have on the pattern of family life? We know that Thomas Cook and Sons, Ltd., world-famous travel agency, started in 1841 when a twenty-three-year-old printer, Thomas Cook, discovered that the new railway company was willing to give him cheaper rates for mass booking of his fellow workmen on a ten-mile outing, from Leicester to Loughborough. Today, Cook and Sons employs more than twenty thousand people and deals with millions of travelers each year. What was the effect of such a development on the "lower classes," often now for the first time exposed to "distant" places and transported to hitherto unknown beach resorts? What ramifications need to be traced into vacation habits, into the growth of hotel accommodations (in nineteenth-century England, for example, the railroads were the largest hotel owners), into the possibilities for "Great Exhibitions" (such as that of 1851) and "World Fairs"? The list is large, and difficult to come to grips with. What Cochran has done here is to give us a good start on an important set of problems.[55]

5. *The Political (Brandfon)*

Brandfon's thesis is simple, but ominous. On one hand, the richest corporation in Mississippi, the Illinois Central, was an outsider, whose objectives were different from those of the State of Mississippi. On the other hand, the people of Mississippi held heightened and false expectations as to the railroad's favorable impact on their lives. The result was what Brandfon calls "alienation for both." Put simply, the lack of realistic awareness of secondary consequences had unfortunate results for both the Illinois Central and the State of Mississippi, results that manifested themselves strikingly in the political arena.

The Illinois Central came to Mississippi because cutthroat competition (as described earlier by Cootner, and Chandler and Sals-

---

[55] An especially important part of the social impact would revolve around the question of new communities formed by the railroad. This would add, for example, the close, fine-grained detail to Cochran's presentation of data concerning large-scale population migration. One might also raise a psychological question such as the threat of impermanence in these towns, bearing in mind the comparable situation of many defense and space communities. One would also want to study problems of community leadership, kinship grouping, and so forth.

bury) on its main lines made it necessary to seek profitable trade elsewhere. Now, while seeking its own ends, profits, the Illinois Central also at first promoted the expansion and well-being of the Yazoo Delta region. As a secondary result, however, it intensified the cleavage between Delta and hill lands, between planter and redneck. Moreover, the very planters who benefited most from the railroad grew suspicious that it was not doing enough to expand cultivation by extending its lines; but the planters themselves were opposed to increased state taxation to accomplish this end out of fear that they would bear the brunt of it. As a result, resented by the rednecks as a foreign and alien power that had strengthened the planter and Negro elements, regarded with distrust by the planters, and with disfavor by all Mississippians whose expectations had been disappointed, the Illinois Central became the object of a tax case brought by the state.

There is no need to go into the complicated details of the case; Brandfon supplies those. It need only be remarked that the tax case showed the Illinois Central to be politically impotent in Mississippi. This gives rise to three reflections: (1) Does this prove the need for more effective professional lobbyists? While the Illinois Central had "friends" in the legislature, they obviously did not turn the tide against popular dissatisfaction. (2) Could any political liaison have succeeded where, in this case, the values of the railroad seemed to run against the values of Mississippi? As Cochran reminded us earlier, the railroads elsewhere seemed to coincide with the generally accepted American values. (3) Did the Illinois Central's centralized management, detailed for us by Chandler and Salsbury, lead to an unnecessary neglect of and aloofness toward local Mississippi needs that could have been amended by a more decentralized bureaucratic railroad structure?

Brandfon's work limits itself to the political impact of the railroad in a single state, Mississippi, as focused around a specific tax issue, although it draws broad implications from the example. Other studies might focus on a larger political scene. Almost all commentators note the role of the railroad as an agent of political unification. Thus, the American federal government was partly inspired to develop its policy of large land grants to the railroads because of the conviction that transcontinental roads would strengthen the ties of union between the Atlantic and Pacific coasts; the Civil War had made evident the problem involved in the South's control of a Mississippi delta through which the Middle West found its chief outlet. So, too, Friedrich List, the German nationalist economist, espoused German railroad development as a means of *Staatsbildung*, and Count Cavour, the future premier of a united Italy, wrote, in 1846, a famous review of the railroads in Italy, extolling their effect on national unification.[56]

Others saw the railroads as a harbinger and exponent of "liberalism" in politics. Thus, while Cardinal Newman recognized this feature in a deprecating way, chiding the liberals for believing that "railroad travelling, ventilation, and drainage and the arts of life, when fully carried on serve to make a population moral and happy,"[57] Macaulay eulogized the railroad because it "not only facilitates the interchange of the various products, of nature and art, but tends to remove national and provincial antipathies, and to bind together all the branches of the great human family."[58]

The fostering of nationalism or liberalism was not the only political result foreseen for the railroad. Still others could see the railroad's form of professional administration as the new model for the direction of man's political existence in a "nonpolitical" way. Thus, the Saint-Simonians, largely responsible for building France's railroad network in the 1850's and 1860's, also postulated an industrial society, run by technicians, scientists, and financiers instead of elected parliamentarians. Eventually, their doctrines found their way into Marxist views about the "withering away of the state" and the subsequent "administration" of society's needs.[59]

Less long-range and utopian, the railroad had a direct political impact, for example, in America, in the formation of the Granger movement and, subsequently, in the rise of Progressivism. In general, however, we need to know more about the railroad's effect on political life in each state, and its effect nationally in redistributing political power. In addition, we should want to link its impact on imagination and on American expectations with other movements in American political thought. The connections are bound to be

[56] See E. J. Hobsbawm, *The Age of Revolution: 1789–1848* (London; Weidenfeld and Nicholson, 1962), p. 179; and Eugene Anderson, Stanley Pincett, Jr., and Donald Ziegler, *Europe in the Nineteenth Century* (Indianapolis; Bobbs-Merrill, 1961), pp. 121–143 (for Cavour's article). On the general subject of *Staatsbildung*, see Eli Heckscher, *Mercantilism*, tr. by Mendel Schapiro (2 vols., London: Allen and Unwin, 1935).

[57] Quoted in Crane Brinton, *The Shaping of Modern Thought* (Englewood Cliffs, N.J.: Prentice-Hall, 1963), p. 180.

[58] Quoted in Walter E. Houghton, *The Victorian Frame of Mind, 1830–1870* (New Haven: Yale University Press, 1957), p. 41.

[59] On the Saint-Simonians, see, for example, Sébastian Charléty, *Histoire du saint-simonisme* (Paris, 1896), and *Henri Comte de Saint-Simon: Selected Writings*, ed. F. M. H. Markham (New York, The Macmillan Company, 1952), pp. xl–xli. For the Marxist view, see Lenin's *State and Revolution*.

subtle, but no less important for that. In any event, the number of state legislators on the payroll of the railroads in America, and the number of MP's (over 120 in 1872) on the directorates of railroads in England suggest that the political impact was not a fictitious one, on that level or on any other.

6. *The Imaginative (L. Marx)*

The sensitive and thoughtful thesis of Marx is that the impact of the railroad on the American imagination must be viewed in terms of a "symbolic landscape" — the pastoral — being penetrated, aggressively and masculinely, by the "iron horse," a symbol of industrialism. In short, the impact on imagination must be studied in imagination's own terms.

According to Marx, the pastoral image or design, stretching back to Virgil and before, represents man's desire to withdraw from society, symbolized by the city, to a rural setting where he may recover his animal and natural self. There is, generally, a political form to this pastoral design, which Marx calls a pastoral ideal. Thus, for example, Jefferson in the eighteenth century projected an ideal of agrarian democracy which was intended to serve as a guide to social policy. Once again, therefore, we must grasp the intimate connection between political life and the imagination that underlies it. The latter supplies the psychological basis for the myths of the former; and as we have seen, myths can take on, as one of their shapes, the lineaments of historical analogy.

In any case, according to Marx, the railroad was perceived as an intrusion of civilization into the pastoral landscape. Thus, Hawthorne, sitting in the woods near Concord, Massachusetts in 1884, remarked: "But, hark! There is the whistle of the locomotive — the long shriek, harsh. . . . It tells a story of busy men, citizens from the hot street, who have come to spend a day in a country village, men of business; in short of all unquietness; and no wonder that it gives such a startling shriek, since it brings the noisy world into the midst of our slumbrous peace." [60] Similar statements, by Thoreau and Emerson, Wordsworth and Thomson, could be cited.

Masculine and aggressive, intrusive and penetrating, the locomotive is seen as the symbol that alienates and estranges man from his earlier world and earlier age: the peaceful, pastoral design. Once and for all, the railroad ends the division of country and city, confirming in the imagination what Cochran has already shown us to be the new reality. It also ends an era, and, as the favorite emblem

[60] This experience is recorded in Hawthorne's notebook.

of progress (the symbol of raised expectations, as we have seen), opens up a new age. As Thomas Arnold remarked, in 1834, on seeing the first train pass through the Rugby countryside, "feudality is gone forever"; to be echoed by Thomas Carlyle's, "Cannot the dullest hear steam-engines clanking around him . . . rapidly enough overturning the whole old system of society." [61]

Destroying one image — the pastoral — and one society — the feudal and rural — the railroad, however, lent itself to a phoenix-like version of the pastoral ideal that Leo Marx calls the middle landscape. In this latter, industrialism was conceived as bringing both equality and the good life to the greatest number of people. For this new development in ideology, poets like Thomson worked out a new literature, integrating the industrial reality with man's psychic needs, and linking the imaginative and political sides of man's nature once again.[62]

Without going into further details, we can see that an understanding of the impact on the imagination of a social invention like the railroad impels us to a study in depth of existing psychological and literary patterns. Marx has done a brilliant job in describing for us the "symbolic landscape" into which the railroad intruded. Other studies might discuss other psychological problems.

One such might concern the fears and phobias associated with the railroad (and, ultimately, with other social inventions). For example, the very opening of the Liverpool and Manchester Railway in 1830, witnessed by crowds upward of fifty thousand, brought with it a tragic and fatal accident to William Huskisson, one of the leaders behind the new railroad.[63] From then on collisions, derailments, and even explosions were constant reminders, colorfully splashed across the front pages of the newspapers (which were themselves distributed by the railroads), of the destructive possibilities of the "iron monster." Superstitious peasants in Central Europe thought the puffing, red-eyed engine was the Devil himself. Even Sigmund Freud had a railroad phobia — much deeper than might be indicated by his placid statement, "though the daily reports [in 1897] of train accidents do not make the father and mother of a family look forward to travelling with any pleasure," — which unveiled itself only through his recapture of a childhood memory:

[61] Quoted in Houghton, *op. cit.*, p. 4.
[62] Cf. Leo Marx, *The Machine in the Garden* (New York: Oxford University Press, 1964), pp. 94–95, 162.
[63] For details, see Robbins, *op. cit.*, p. 14.

"At the age of three I passed through the station when we moved from Freiberg to Leipzig, and the gas jets, which were the first I had seen, reminded me of souls burning in hell. I know something of the context here. The anxiety about travel which I have had to overcome is also bound up with it." [64]

On another level, one might study the impact of the railroad's creation of its own environment on people's ideas of the possible, to revert to a theme taken up in the chapter by Hughes; or the railroad's "annihilation of time and space," to use de Tocqueville's rephrasing of a line by Alexander Pope: what impact did this have on men's imagination? [65]

In some ways the most difficult to trace and estimate, the railroad's impact on the imagination seems almost to be the most fundamental. Touching the most hidden, and therefore most difficult to fathom, sources of our psyches, the impact on imagination underlies and conditions all of our other reactions. On the surface, as Cochran has pointed out, the coming of the railroad, and inferentially of industrialism, was conditioned by existing American values, which were then reinforced. In the depths, Marx suggests to us, the coming of the railroad can only be understood in terms of an existing "symbolic landscape," across which the railroad twisted and turned, leaving psychic scars but also opening up new, imaginative possibilities. What, one asks, will be the impact of other such social inventions?

[64] *The Origins of Psychoanalysis. Letters, Drafts and Notes to Wilhelm Fliess*, ed. Marie Bonaparte, Anna Freud, and Ernst Kris (Garden City, N.Y.: Doubleday, 1957), pp. 217, 240.
[65] Independently of the historical analogy on railroads, the Committee on Space commissioned an interesting study by David Lowenthal on "Visions of Space: Some Perspectives on the Impact of Space Exploration." This paper does offer some suggestive insights on the "annihilation of time and space" that has been taking place since the fifteenth century. It gives a good, general summary of ideas and attitudes held over the last four to five hundred years on these matters, and presents a most useful bibliographical summary on the subject. Lowenthal divides his material under four categories of discovery and four aspects of world views. The four categories of discovery are (1) geographical, (2) technological invention, (3) sensate extensions (as with the telescope), and (4) ideas; and the four aspects of world views are (1) space and spatial order, (2) time and temporal order, (3) matter — which includes gravity — and energy, and (4) life and man's place in the natural, spiritual, and social orders. Because of the paper's wide range over time and space, following "visions of space" rather than studying a bounded, interrelated social complex in detail and then the impact of a single social invention thereon, it does not fit the conditions for a historical analogy that I have laid down earlier; therefore, it would be unwise to attempt to fit it in as an integral part of the present study. Nevertheless, as a suggestive overview of the four categories of discovery and world views it should be consulted.

7. *Some Generalizations*

In the discussion of historical analogy, I stated that the highest aim, scientifically, would be the production of generalizations that could then be applied to further materials. This is not to say that hunches or intuitions derived from the study of a historical analogy are to be looked down upon; they are valuable in themselves, and may be all that we can secure. Indeed, the attempts at generalizations that follow cannot pretend to be rigorously and narrowly formulated. In fact, they may be considered only a halfway step from generalities to generalizations. In any case, they are, briefly, as follows:

A. *All social inventions are part and parcel of a complex — and have complex results. Thus, they must be studied in multivariate fashion.* While this may seem like stating the obvious (as does the statement that the earth revolves around the sun — after it is made), in practice, it is not. Let me give a concrete and simple example from railroad history. Isambard Kingdom Brunel was as great an engineer as his rival, Robert Stephenson (George Stephenson's son). Brunel persuaded the Great Western Railway, in 1835, to adopt the broad gauge of seven feet for their line. As Robbins tells the story in *The Railway Age*:

> There were, and still are, most compelling technical arguments in favor of it; but one thing ought to have been decisive to the contrary — it was already too late. George Stephenson, with his usual heavy sagacity, had advised the Leicester & Swannington directors so six years before. The idea of the broad gauge was not impracticable — indeed, it was eminently practical; but in the long run the broad gauge could not keep a whole sector of the country to itself, and the flaw was fatal. Robert Stephenson, with his more earth-bound understanding, saw this clearly; Brunel would never admit it. The broad gauge was a failure not of technique but of administrative imagination.[66]

To move from this particular, limited case, as we ourselves have done via the Chandler and Salsbury chapter, we can see that the "administrative imagination" must encompass a complex, integrated system of relations. And so must the student of social impact. Indeed, it is not only the administrative imagination, but the technological, economic, social, political, and so forth that must be understood in detail. In short, this first generalization should serve as a warning to anyone, scholar, publicist, or policy maker, to be cautious about embracing simple views as to the nature and impact

[66] Robbins, *op. cit.*, pp. 23–24.

of a social invention. Moreover, it reminds us that there is no substitute for detailed study as to the nature of that complex effect.

B. *No social invention can have an overwhelming and uniquely determining economic impact, and this is so partly because no completely new innovation is possible in reference to any set of economic objectives.* The details supporting this generalization are to be found in the chapters by Cootner and Fogel. I shall only add here that I find this generalization consoling — the "cataclysmic," single invention does not appeal here any more than in geology or biology. This generalization, moreover, appears to be a necessary corollary of the complexity of social relations assumed in the previous generalization.

C. *All social inventions will aid some areas and developments, but will blight others.* In short, an asset and liability bookkeeping is necessary to the study of any impact. Once again, this follows from the complexity assumed in the first generalization. It is also a fundamental part of the analysis, for example, as stated by Cootner and Fogel, that underlies the second generalization.

D. *All social inventions develop in stages, and have different effects during different parts of their development.* For example, as Chandler and Salsbury have pointed out, the railroad's impact on managerial practices involved at first professional engineers, and then later on financiers at the top level of the bureaucracy; and varied from a centralized to a decentralized, and back to a centralized administrative structure. So, too, as Cootner and Fogel have shown us, railroad investment came in phases, and in different areas — industrial and agricultural — with the maximum impact on the economy about fifty years after the so-called technological invention itself. The caution here is that the impact of a social invention is not only complex in terms of an assumed static social relationship but in terms of dynamic time as well.

E. *All social inventions take place in terms of a national "style," which strongly affects both their emergence and their impact.* Thus, as Hughes points out, American railroad technology was shaped by the great distances of this country and the need to open up quickly the resources of the interior, and thus to opt for temporary expedients and to compromise with nature. Chandler and Salsbury remind us that American railroads, faced with distances unknown to their British counterparts, turned to decentralized administration, with later repercussions for all of American industry. Cochran has talked of American values, and the encouragement that they afforded to railroad builders, and Marx has investigated the psychological-political meaning of the American version of the pastoral landscape upon which the railroad seemed to break so abruptly. Again to revert to the first generalization, we can see that the complexity of which we have spoken is characterized and unified by a particular and identifiable "style" or "way" of arranging social relations.

Stated in the abstract, these generalizations may appear both obvious and vague. Certainly, if possible, they need to be tightened, and their boundary conditions more clearly stated.[67] If this is impossible, then we shall have to be satisfied with them as heuristic devices, calculated to lead us aright in our detailed empirical researches. Whether they are acceptable generalizations or merely pragmatic guides, I believe that they can be overlooked in the analogical study of historical social inventions only at our own peril.

III. The Space Program as a Social Invention

An analogy should consist of at least two elements. In the case before us, comparing the impact of the railroad and the space program, the second element is largely unknown; indeed, we have been using the railroad in part as a "device of anticipation." However, exactly because the impact of the space program is as yet largely unknown — a separate study in itself — I can only make a few comments on it; anything else would be foolhardy. My plan here is to notice some of the prophecies made about the two social inventions; to compare their putative purposes; to discuss briefly some questions of philosophical import; to comment on the technological-scientific aspects of the space program; and, lastly, to recommend, on the basis of our research into the railroad impact, some specific research topics (other than those already recommended in the individual railroad papers themselves) concerning the space program.

Prophecies

When the railroad first loomed on society's horizon, the prophets went quickly to work. For example, the British magazine, *Quarterly Review*, in 1825 wrote: "Can anything be more palpably ridiculous than the prospect held out of locomotives travelling twice as fast as stage coaches. We should as soon expect the people of Woolwich to suffer themselves to be fired off upon one of Congreve's Rockets

---

[67] For example, the second generalization may apply only to modern, industrialized society, and not to primitive societies (where, for example, a new grain crop may change drastically not only the economy but the entire culture).

as trust themselves to the mercy of such a machine going at such a rate." A booklet of 1827 dismissed the possibility of a locomotive traveling at sixty miles an hour as "the pure effulgence of an untaught mind and of limited talent." [68] Against the pessimists, however, the optimists ranged themselves with equally strong prophecies. Thus, Macaulay predicted that in the twentieth century there would be no more highways or streets, for everything would move on rails.[69]

By 1867, half as prophecy and half as summary, *The North American Review* could contemplate the effects of the railroad:

"With perhaps two exceptions," it said, the railroad was "the most tremendous and far-reaching engine of social revolution which has ever either blessed or cursed the earth." It was this innovation which "made our century different from all others, — a century of greater growth, of more rapid development." The railroad, the distinguished journal held, had peopled the "wilds of America" and turned the "very Arabs of civilization" into "substantial communities." It cosmopolitanized the nation; removed the old distinctions between classes, changing both dress and manners; "abolished" the Mississippi River; crushed the Southern rebellion; made "the grass grow in once busy streets of small commercial centers like Nantucket, Salem, and Charleston"; robbed "New Orleans of that monopoly of wealth which the Mississippi River once promised to pour into her lap," and simultaneously turned New York into "an overgrown monster." The "iron arms" of the railroad, the *Review* concluded, "have been stretched out in every direction; nothing has escaped their reach, and the most firmly established institutions of man have proved under their touch as plastic as clay." [70]

Unlike *The North American Review*, however, we have attempted in our specific studies of the railroad to weigh, not the prophecies, but the real effects of the railroad.

One "lesson," however, to be drawn here is that *any* social invention is bound to be surrounded by prophecies, pro and con. Their value is usually very little, except as statements of mythical aspiration or of model-like intention — a significant exception, of course. Thus, even before the "Space Age" itself, men like Jules Verne were prophesying the wonders to come, while the novelist Winwood Reade in the *Martyrdom of Man* (1872) expressed his ecstatic belief that, after various earthly gains,

Then, the earth being small, mankind will migrate into space and will cross the airless Saharas which separate planet from planet, and sun

[68] Quoted in Andrews, *op. cit.*, pp. 15–16.
[69] Brinton, *op. cit.*, p. 153.
[70] As summarized in Fogel, *op. cit.*, Chapter 1, pp. 3–4.

from sun. The earth will become a Holy Land which will be visited by pilgrims from all the quarters of the universe. Finally, men will master the forces of Nature; they will become themselves architects of systems, manufacturers of worlds. Man then will be perfect; he will then be a Creator; he will therefore be what the vulgar worship as a God.[71]

In our time the claims have generally been a little less God-like. Thus, Robert Jastrow and Homer E. Newell suggest that the space program will bring about "the extension of man's control over his physical environment" (a sober enough estimate of NASA's primary mission), although they then go on to claim that "the science which we do in space provides the equivalent of the gold and spices recovered from earlier voyages of exploration." They conclude that the specific value of space exploration will be "The benefits of basic research, economically valuable applications of satellites, contributions to industrial technology, a general stimulus to education and to the younger generation, and the strengthening of our international position by our acceptance of leadership in a historic enterprise."[72] The economic benefits, in terms of "spin-off," have been especially touted. Thus, Toby Freedman, Director, Life Sciences, North American Aviation, Inc., appeals to "one of the great laws of history: the pioneer shall have his discovery. It may not be what he was looking for. But it may be better." Then Freedman claims at least a hundred pages of notes on spin-off products from the space program. He ends by announcing that in his own field of "medical miracles" contributions exist "that to my mind have already paid back the cost."[73]

While as far back as the seventeenth century, men like John Donne and Francis Harding foresaw the moon as a dumpheap for undesirables (Jesuits, according to Donne) — a sort of penal settlement à la America — their modern American successors talk of the planets as the solution for the problem of overpopulation. "If you want to look ahead a hundred years or more," says Murray Zelikoff of Geophysics Corporation, "I think the real purpose of the space

[71] Quoted in Houghton, *op. cit.*, p. 36.
[72] Robert Jastrow and Homer E. Newell, "Why Land on the Moon?" *Atlantic Monthly* (August 1963), pp. 42–43, 45.
[73] *This Week Magazine* (March 8, 1964), pp. 12–13. On the other hand, Joseph Goldsen prophesies in a more sober vein that "unless the government underwrites the risks of adopting new technologies, which is not expected nor advised, the 'spin-off' consequences may be evident only in terms of very long time spans, say 25 years or more." *American Behavioral Scientist*, VI, No. 7 (March 1963), 6.

effort is to colonize the planets." [74] In a different area, another glowing view, this time concerning the problem of education, is advanced by Erik Bergaust: "Fifty years from now? Who knows, perhaps we will terminate the use of the title *doctor* — because everyone will have at least a Ph.D. degree. That might well become a typical result of our current Space Age brainpower drive." [75]

The optimists, of course, have not had it all their own way. Thus, Edwin Diamond likens the space race between America and Russia to the "potlatch" ceremony of the Kwakiutl Indians, where "two clans would gather around a fire and their chiefs would vie with each other in throwing the clan's most valued property into the fire," while Lewis Mumford apparently sees a resemblance between the Space Age and the pyramid age, commenting, dryly, that the pyramid age, with its enormous wasteful expenditures, came to an abrupt end.[76]

Without citing any more examples of prophecies, pro or con, concerning space or the railroads, one more "lesson" ought to be clear. Both optimists and pessimists never waver in their violation of our third generalization: "All social inventions will aid some areas and developments, but will blight others."

Purposes

The line between prophecy and statements of purpose is often thin. With this underlined, it ought next to be said that initial or primary purposes often give way to others, and that even these new purposes may yield in importance and primacy to the secondary consequences that they in turn produce. Thus, the railroad's primary purpose was to haul coal from the mines; this quickly gave way to the primary purpose of hauling passengers and bulky goods other than coal. Then, as we have seen, secondary consequences, such as the achievement of national unity, or the fulfillment of heightened expectations, came partially to supersede in importance the earlier primary goals of the railroad builders.

[74] Quoted in *The Space Industry: America's Newest Giant*, by the editors of *Fortune* (Englewood Cliffs, N.J.: Prentice-Hall, 1962), p. 97. Cf. Lowenthal, *op. cit.*, p. 68.

[75] Erik Bergaust, *The Next Fifty Years in Space* (New York: Macmillan, 1964), p. 57.

[76] Edwin Diamond, "The Rites of Spring," *Bulletin of the Atomic Scientists*, XIX, No. 5 (May 1963), p. 26; Mumford's statement is contained in "An Analysis and Summary of Responses from Fellows of the American Academy of Arts and Sciences Concerning the Space Program as a Subject for Study and Thought," unpublished document of the Committee on Space Efforts and Society of the American Academy of Arts and Sciences, ed. by Geno Ballotti.

As for the space program, the official statement of its purposes is put forth in the National Aeronautics and Space Act of 1958. There it is stated that the objectives are

(1) to conduct the scientific exploration of space for the United States (a function of NASA);
(2) to begin the exploration of space and the solar system by man himself (a function of NASA);
(3) to apply space science and technology to the development of earth satellites for peaceful purposes to promote human welfare (a function of NASA);
(4) and, to apply space science and technology to military purposes for national defense and security (a responsibility of the DOD).[77]

Obviously, however, an undeclared purpose lurked behind this official statement. On October 4, 1957 the first Soviet satellite was launched. NASA's purpose, therefore, seen against this event, was to recover for America the lead in space science and engineering and with it the prestige of being the most advanced industrial society, which seemingly had fallen to the Russians; coincidentally, of course, NASA's purpose was also to rectify any shift in the balance of military power that might have occurred.[78] Soviet achievements in February and March of 1961, the launching of four spacecraft, each weighing ten thousand pounds or more, and on April 12, 1961, the successful orbiting and recovery of Major Gagarin, may be recognized as also having prompted the speed-up of the Apollo project for manned lunar landings.

On another level, however, the answer to the question, "Why go to the moon?" (and thus into space generally) was given by President Kennedy on September 12, 1962, when he quoted the words of the British explorer, George Mallory, who was once asked why he wanted to climb Mount Everest: "Because it's there."[79] At first glance perhaps trivial or facetious, this response gains support from the high respect afforded the instinct of exploration by biologists and physiologists. Thus, for example, S. A. Barnett remarks that "in nature, exploratory behavior gives an animal information about its surroundings which can be used later — in a search for food, flight from danger and so on. It makes possible a process called latent learning because its effects become evident only on a subsequent

[77] *Historical Origins of the National Aeronautics and Space Administration* (Washington, D. C., 1958), p. 5.
[78] For an examination of this thesis, see Enid Curtis Bok, "Making American Space Policy: (1) the Establishment of NASA," working paper, School of Industrial Management, Massachusetts Institute of Technology, January 1963.
[79] Quoted in Diamond, "The Rites of Spring," *op. cit.*, p. 27.

occasion." [80] In this same vein, some have claimed that space must be explored as part of the "ineluctable progress of science." (Thus, the director of the Office of Space Sciences of NASA states flatly that the primary object of the space-science program is to extend human knowledge, and that it is fundamentally a basic research effort.) [81] Other statements by President Kennedy are that "We go into space because whatever mankind *must undertake* [my italics], free men must fully share," and, "We sail on this new sea because there is new knowledge to be gained, and new rights to be won, and they must be won and used for the progress of all people." [82]

Clearly, there are many stated purposes for the space program. As is to be expected, different people have different goals in mind. So, too, some of the stated purposes have an exhortatory, as well as a prophetic or declaratory function. The point to make here, however, is that the analogy of the railroad suggests to us that the stated purposes do not provide a good guide to the actual impact itself. Significant as expressing expectations, or imaginative or mythical longings, the stated primary purposes in fact give way to unexpected secondary consequences, which then often turn out to have the greatest impact on society. The problem facing us today, and the one with which this study is concerned, is how to achieve prior knowledge of these secondary consequences, and thus to be in a position more adequately to contend with them. The "lesson" of our historical analogy with the railroad seems to be that we must not be misled by the rhetoric of supposed primary purposes.

Philosophic Impact

Certainly, the farthest remove from an intended primary aim of the space program is a philosophical impact. Yet, in the long run, this may be one of the most significant effects. While it could be treated under the heading of "imagination," I think it best to say a few words about it separately, and then proceed to the rest of our impact topics.

Actually, the philosophic aspects of the space program go back to earlier developments. Thus, as we have seen with Hughes, the railroad, symbolizing modern technology, signified the conquest of natural conditions and the *creation* by man of a *self-made environment*. In an important sense, of course, this was simply a continua-

---

[80] S. A. Barnett, " 'Instinct'," *Daedalus*, 92, No. 3 (Summer, 1963), 568.
[81] N. J. Berrill, "Our Gamble in Space," *Atlantic Monthly* (August 1963), p. 36.
[82] Quoted in *Historical Origins, op. cit.*, pp. 15–16.

tion of man's creation of the city, another artificial environment. But it did add an extra dimension, which made, more or less, for a qualitative distinction. Now, with the space program, we seem to have taken another giant step. Thus, the program for the Fourth National Conference and Space Exposition, 1964, discussing the various spacecraft, states that "man's increasing mastery of the new element enables him to take his environment along with him." [83] Moreover, the spacecraft appears to offer the only known means of creating a weightless environment (and, as one commentator remarks, "Scientists believe that fundamentally new knowledge of the nature of life will be gained by observing human and other biological life in such a space laboratory").[84] Put another way, the new research tools of the Space Age have "made it possible, for the first time, for man to escape from this environment [the earth] and look directly into naked space."[85] Man, himself, becomes part of this self-created environment, as space medicine deals with men in stress in more or less the same way that space engineering deals with materials in stress.[86] In short, the development that Hughes saw in relation to railroad technology has proceeded apace.

What are man's reactions to this Faust-like development? First, we must recall that the railroad, as a symbol of industrialism, had already destroyed or altered the pastoral landscape. Man, therefore, no longer conceives of nature as something given, maternal in its existence, on which he reclines; instead, nature is something to be obliterated so that a second creation, to use Hughes' phrase, may occur. In place of the pastoral image, we now have a new image, of space, which is inhabited by a spacecraft itself freed from the environment of the earth.

Next, we must realize that many features of such an environment have previously been considered a form of torture or deprivation. Thus, Lewis Mumford views interplanetary travel as the culmination of modern industrial society's drive toward sensual starvation: "Under such conditions, life would again narrow down to the physiological functions of breathing, eating, and excretion. . . . By comparison, the Egyptian cult of the dead was overflowing with vitality; from a mummy in his tomb one can still gather more of

[83] It adds, "and will finally provide lunar travel in shirtsleeve comfort."
[84] Franklin A. Lindsay, "The Costs and the Choices," *Atlantic Monthly* (August 1963), p. 52.
[85] Quoted in Jessup and Taubenfeld, *op. cit.*, pp. 110–111.
[86] Cf. Donald N. Michael, *Proposed Studies on the Implications of Peaceful Space Activities for Human Affairs* (Washington, D. C., 1961), p. 16.

the attributes of a full human being than from a spaceman."[87] While, probably, most people today would reject Mumford's forceful condemnation — after all, an age which knows submarines and bathyspheres will not balk at spacecraft — it does remind us of a long and important trend in speculation, wherein man's increasingly artificial life is seen as a condition of estrangement and alienation from himself.

Indeed, the philosopher, Hannah Arendt, seeks to take up this critique. She sees the space effort as allowing man to realize in fact what Copernicus and others had postulated in theory: to stand in the sun and overlook the planets. So, too, Einstein's imagined "observer freely poised in space" now becomes possible in actuality, as does the eventual realization of Einstein's "twin paradox" (which hypothesizes that "a twin brother who takes off on a space journey in which he travels at a sizable fraction of the speed of light would return to find his earthbound twin either older than he or little more than a dim recollection in the memory of his descendants").[88] Faced with these possibilities, Arendt's comment is dire, indeed:

We have come to our present capacity to 'conquer space' through our new ability to handle nature from a point in the universe outside the earth. For this is what we actually do when we release energy processes that ordinarily go on only in the sun, or attempt to initiate in a test tube the processes of cosmic evolution, or build machines for the production and control of energies unknown in the household of earthly nature. Without as yet actually occupying the point where Archimedes had wished to stand, we have found a way to act on the earth as though we disposed of terrestrial nature from outside, from the point of Einstein's 'observer freely poised in space.' If we look down from this point upon what is going on on earth and upon the various activities of men, that is, if we apply the Archimedean point to ourselves, then these activities will indeed appear to ourselves as no more than 'overt behavior,' which we can

---

[87] Lewis Mumford, *The Transformation of Man* (New York: Harper, 1956), pp. 175–176.

[88] Hannah Arendt, "Man's Conquest of Space," *American Scholar* (Autumn, 1963), p. 535. Cf. J. Bronowski, "The Clock Paradox," *Scientific American, 208,* No. 2 (February 1963), 134–144. I find Edward Teller's comment on the "twin paradox" possibility especially amusing. "A new race will live which we might consider as a strange and horrible new species, but which in reality will be far superior to us and much better. And what they will do with me — a specimen from an old, fabulous, unreasonable, extinct race — is obvious. I will be put in a zoo." *Peacetime Uses of Outer Space,* ed. Simon Ramo (New York: McGraw-Hill, 1961), p. 271. If I may be allowed a facetious comment à la Mumford: after many days and weeks in a spacecraft, a cage in a zoo may seem a fairly "natural" environment.

study with the same methods we use to study the behavior of rats. Seen from a sufficient distance, the cars in which we travel and which we know we built ourselves will look as though they were, as Heisenberg once put it, 'as inescapable a part of ourselves as the snail's shell is to its occupant.' All our pride in what we can do will disappear into some kind of mutation of the human race; the whole of technology, seen from this point, in fact no longer appears 'as the result of a conscious human effort to extend man's material powers, but rather as a large-scale biological process.' Under these circumstances, speech and everyday language would indeed be no longer a meaningful utterance that transcends behavior even if it only expresses it, and it would much better be replaced by the extreme and in itself meaningless formalism of mathematical signs.

The conquest of space and the science that made it possible have come perilously close to this point. If they ever should reach it in earnest, the stature of man would not simply be lowered by all standards we know of, it would have been destroyed.[89]

Whether or not one shares this Cassandra-like reaction to man's presence in space — and I happen not to — it is clear that the space program has and will have a growing philosophical impact.[90] Man's view of himself and of his own nature, as well as of Nature, is bound to be affected by his increasing ability to create a man-made environment for a man-made man.

The Technological

Aside from stressing again the value of looking at space technology in Hughes' terms of the conquest of a "natural frontier," I have only one or two brief comments to make. The first is that technology and technological diffusion seem to be the most obvious features of the space program. We *see* the space missile on our TV sets, and we shall probably soon *enjoy* some of the "civilian" goods produced with the new technologies, such as miniature electronics or heat-resistant materials. Improved weather forecasting — a technological spin-off — will shortly be with us, and a host of other "unintended" items. The only caution in studying the impact of the space program in terms of technology is our third generalization. Thus, many of these technological developments might have occurred, perhaps more gradually, without the space program; the costs of development might be economically undesirable; and other developments might have been blighted by the emphasis on the

[89] Arendt, *op. cit.*, p. 539.
[90] See my article, "The Idea of Progress," *Daedalus*, 92 (Summer, 1963), for some comments on "man's condition" that differ from the view expressed by Arendt.

space program (compare various ongoing studies of engineering and scientific manpower and their utilization); and so forth. Without diminishing the importance of the contribution of the space program to technology, as a social invention we must study its impact in this area in a balanced and sophisticated fashion.

My second comment is that there exists a significant difference between railroad and space technology in that the latter powerfully lends itself to the development of science while the former did not (and this ultimately entails a very different secondary effect in so far as economic life is concerned). True, the railroad probably contributed in some measure to the formulation of the second law of thermodynamics — as P. B. Medawar says, the latter "was in due course taken out of its native environment, here the pithead and the railway workshop, and generalized."[91] And, trivially, the railroads' accidental excavations of fossils may have inspired some naturalists. The space program, on the other hand, has as one of its prime aims the extension of scientific knowledge, and about 15 per cent of NASA's funds are directly assigned for this purpose.[92] The rationale for much of the technological "hardware," in fact, is to obtain the data on which scientific theories about space can be established. So significant is the difference of the two social inventions — it is also a measure of the increasing role of scientific research in our society — that a separate consideration of the scientific impact of the space program, unnecessary for the railroad, is here requisite.

The Scientific

The scientific consequences — both intended and unintended — of the space program will probably be of the highest importance. Three main areas of investigation are already clearly delineated (though in part, of course, they overlap). The first concerns knowledge about the earth itself. To quote one source,

Much new information has already been obtained, for example, about the actual shape of the earth, which seems to resemble more a pear than the traditionally taught flattened orange. This knowledge permits more accurate calculation of distances on Earth and more accurate map-making. Information will be obtained on such subjects as whether or not the earth is heating up or cooling down, the nature of the atmospheres of the earth and other heavenly bodies, the nature and extent of electric, magnetic,

[91] P. B. Medawar, "Onwards from Spencer," *Encounter,* XXI, No. 3 (September 1963), 39. For details of this development, see A. R. Ubbelohde, *Man and Energy* (Baltimore, Md.: Penguin Books, 1963), pp. 125 ff.
[92] Berrill, "Our Gamble in Space," *op. cit.,* p. 36.

and gravitational forces, and the like. One of the most striking discoveries already made, for example, is that of a belt of high-energy radiation extending upward from a height of several hundred miles and substantially disappearing beyond 17,000 miles from Earth.[93]

A second concerns the nature of space itself. Until recently viewed as a serene and empty void (although Descartes' vortex theory had assumed otherwise), "today physicists know that near space is filled with radiation belts, solar winds, and great plasma clouds."[94] Also, once outside the earth's atmosphere, whether ourselves or with instruments, we are able far better to plumb the depths of galactic and intergalactic space, and to understand their features and characteristics.

A third concerns the possibility of extraterrestrial life (even at a very low level). As one biologist glowingly puts it, "The scientific question at stake is the most exciting, challenging, and profound issue of the whole naturalistic movement characteristic of Western thought for three hundred years, ranking with the revolutionary impact of Darwin and Copernicus."[95] In point of fact, Mars seems to hold the only hope for such extraterrestrial life. However, "if Mars can be shown to possess any sort of indigenous life at all, we may feel assured that living organisms originate wherever circumstances permit, as a natural and inevitable end of universal matter."[96]

Now, exactly what social impact knowledge in these areas will have is extremely difficult even to hint at. Jastrow and Newell seem bold when they foresee that the new field of space science will revive "the spirit of catholicity in science" as "an important accompaniment to space research." On the other hand, N. J. Berrill, a biologist, warning against the dangers of contaminating outer space, or of being contaminated, turns naturally to a historical analogy: "Sending a man to Mars, if he is to become in any way exposed, is as big a threat to the life of that planet as we can make, and leaving him return home as a carrier of exotic microbes would be just as dangerous to us. The devastating impact of the white man's germs on Pacific islanders and Eskimos during the nineteenth century should not be forgotten."[97]

Berrill's analogy is, of course, largely illustrative. As for our rail-

---

[93] Jessup and Taubenfeld, *op. cit.*, p. 226.
[94] Diamond, "The Rites of Spring," *op. cit.*, p. 27.
[95] Berrill, *op. cit.*, p. 36.
[96] *Ibid.*, p. 38. See the entire article for details of this fascinating subject. See, also, Edward Teller, "Outer Space Travel — What Is and Is Not Possible?" *Peacetime Uses of Outer Space*, ed. Ramo, pp. 264–266.
[97] Berrill, *op. cit.*, pp. 45, 39.

road analogy, in this area it offers almost nothing, either illustrative or heuristic (although, under philosophic impact I have tried to put down a few reflections). Obviously, only a detailed example from the history of science, preceded by some sort of theoretical consideration of the historiographic problem involved, would be of service in trying to estimate the impact of the space program on science, and of the resultant scientific knowledge on society. It need hardly be stated that an attempt at such an effort, here, is beyond my intention or capabilities.

Other Areas

Only specific research, comparable to that on the railroad impact, into such areas as the economic, managerial, social, political, and intellectual effects of the space program can produce valid knowledge rather than glittering generalities. Statements on the economic effect, such as that "it is certainly safe to say that the introduction of a new industry — the aero-space industry — on the American scene is having a healthy and wholesome effect on the national economy," or that "our space outlays will yield $2 return for $1 invested," may be fine propaganda, but they are not adequate as analysis.[98] Obviously, without detailed studies, further comments in these areas lie beyond the intention of the present report.

Instead, I should like to suggest, on the basis of our research into the railroad impact, some possible research topics concerning the space impact. These are in addition, although sometimes there may be a slight overlap, to the recommendations made in the various chapters on the railroads themselves.

A. *Economic.* 1. Cootner states that "projection of prospective impacts of innovation must be developed with careful regard for the relative economies of production," and Hughes has called our attention, too, to the interconnectedness of technology and economics. What happens, however, when market requirements and "relative economies of production" are either eliminated as factors, or else diminished in importance? Obviously, therefore, there are enormous and novel difficulties in analyzing the economic aspect of the space program. At the over-all level, space economics can be regarded, unlike the railroad, as noncompetitive. While the problem of allocation of limited resources, of course, still exists,[99] de-

[98] Bergaust, *op. cit.*, pp. 9, 13.
[99] Cf. Raymond Bauer, "What are the Social Implications of our Space Program?" Address to the Public Relations Institute, Cambridge, Mass., July 29, 1963, p. 2.

cisions have tended to be administrative rather than market ones. On a lower level, however, there may be market-type competition among the support industries that must appeal to NASA as their customer. Moreover, on closer examination, the fact of governmental financial support of the space program appears to differ only in degree from the railroad case: even before the Civil War, Congress dispersed over 100,000,000 acres of the public domain, while state and local governments provided in cash or credit an additional 30 per cent of the total capitalization of railroads.[100] All in all, then, it is clear that factors such as those cited, and many more, will have to be dealt with in a refined way. In short, an important research topic will raise the question: what is the impact on both technology and the economy brought about by the space program's creation of a seemingly new relationship (or is it a matter of degree?) between the two?

2. In the case of the railroads, we saw that the secondary economic impact was not necessarily on domestic industries (for example, whereas the demand for engines tended to be satisfied in the United States, the demand for iron and steel led to imports from England). Our questions, therefore, are what has been, and what may be the secondary effect of the American space program on the industries of other countries? Will part of the supposed benefits of the space program to American economic life actually flow to other countries?

B. *Managerial.* What have been the critical decisions in the space program as to the administrative setup, and what have been the results of these decisions? What critical decisions may lie ahead? (The Chandler and Salsbury proposals, pp. 160–162, are so carefully worked out that I shall refrain from any further suggestions in this area.)

C. *Social.* 1. According to one scholar, Derek Price, the number of scientists has been doubling every fifteen years since about 1750. Moreover, today, there are probably more scientists living than have existed in the whole of man's past.[101] While even in the United States, however, the present number only amounts to about five hundred thousand,[102] it is clear that the scientific community is growing ever larger and more important. What consequences does

---

[100] Fogel, *op. cit.*, Chapter 1, p. 5.
[101] Derek J. DeSolla Price, *Science Since Babylon* (New Haven: Yale University Press, 1961), Chapter 5.
[102] Profiles of Manpower in Science and Technology (Washington, D. C.: National Science Foundation, 1963), p. 7.

this have for society at large? A number of lines of inquiry open up.

(a) The space centers exist as almost unique concentrations of engineers and scientists (with Oak Ridge, Tennessee, as a precursor: was it studied ahead of time as an example of problems that might be encountered at Huntsville, or Cape Kennedy?). Do we have here, in communities where probably half the residents are "scientists," a preview of our future society at large? Would a study of their attitudes and institutions serve as a microcosm of the world to come? Huntsville, for example, is said to have no working class per se, no unions, and no "class conflict"; instead, there is only vertical conflict, between young and old. Is this the shape of the future?[103]

(b) Does the "team" of Dr. Wernher von Braun and his German rocket experts point to a special type of "employee" developing among scientists (that is, not serving so much for pay as for the pure, professional exercise of one's craft or science under any patron)? What effect, if any, does this have on the scientific "community"? What "style" of science does it bring with it?

(c) Will the increasing importance of technical and scientific skill drastically affect the status system as well as the power system in America? As Daniel Bell remarks, "The changes in the structure of power created by the erosion of property as a significant variable and the striking emergence of technical skill as the necessary condition for holding high position have already increased . . . the sense of 'dispossession' — the loss of place and power in the society — that is felt by the older business and professional members of middle-class America."[104] Is this analysis borne out by the microcosm of the new space communities?

2. Has achievement rather than profit become the goal of space-center operations, and what long-range social effect might this have? What motivates the new "managerial class"? And what does this signify about the sort of "manpower" that will be attracted to future programs?

3. Some people saw the railroad, along with industrial society in general, as violating God's intentions. What is the *image* held

---

[103] See *The Space Industry, op. cit.*, Chapter 4, for a description of Huntsville, and cf. Chandler and Salsbury on the railroad's role in the introduction of labor unions. Also, see the various reports to the Committee on Space by Peter C. Dodd.

[104] Cf. Chandler and Salsbury, on the separation of ownership and control, and see Daniel Bell, "The Ethnic Group," a book review in *Commentary*, 37, No. 1 (January 1964), 76.

by religious-minded individuals as to the significance in this area of the space program (I would guess that almost all religious-minded adults would pooh-pooh fears of attrition to religion caused by the space probes)? What is the reality, however, of that effect, as measured, for instance, by the immediate and unconscious reactions of children?

4. Cochran claims that American values contributed to and were reinforced by the coming of the railroad. What is the situation in regard to the space program?

5. What is the American "style" of operating the space program (technologically, administratively, etc.), as contrasted, say, with the Russian?

D. Political. 1. What form have NASA's attempts at lobbying or influencing Congress and/or the executive branch taken? Has this had any special effect on American political life?

2. Has the space program had any effect on American courts and the law? Can any such effects for the future be foreseen? [105]

3. What aims, beyond its primary goals, does NASA have in the local areas where it has introduced itself (e.g., Huntsville, or Michaud)? How realistic are these (for example, does expanded education involve changed race relations)? How does the belief that NASA automatically must create general prosperity for everybody in the area, in other words, heightened expectations, operate in the face of contrary results?

E. Imaginative. 1. To what extent is the space effort seen as an "intrusion" into the "pastoral landscape" described by Leo Marx? Is the whine of the jet engine the new form of the shriek of the railroad whistle? Or is it that once the railroad has shattered the pastoral image, this particular experience is no longer possible? Instead, will space flights somehow take on the quality of being redemptive journeys away from society in the direction of nature? [106]

2. As a corollary, what effect will the "space adventure" have on man's conception of nature? Will nature in fact be redefined, not in the pastoral sense, but in the sense of a man-created entity, as Hughes has suggested?

3. Leo Marx talks at one place of "the rhetoric of the technological sublime," [107] that is, of the favorable hyperbole used about the coming of the railroads, which were thus treated as the symbol of

---

[105] Cf. Cochran, p. 174, for the railroad effect on law. See, too, Jessup and Taubenfeld, *op. cit., passim.*

[106] Cf. Marx, *op. cit.,* p. 69, for an analysis of Shakespeare's play, *The Tempest,* in these latter terms.

[107] *Ibid.,* p. 214.

progressivism. Yet, as he points out, this language also covertly carried with it the fears and doubts of its speakers: "iron monsters" and "dragons of mighty power" implicitly associate the new social invention with the destructive and the repulsive. Only a subtle analysis can reveal the ambivalence of response to the new element. What, in the case of the space program, is its "rhetoric of the technological sublime," and what does this rhetoric reveal about our responses to the "space adventure"?

4. Eighteenth-century poets, like James Thomson, sought to adapt pastoral literature to the new industrial reality. What efforts are being made by literary men today to integrate the space experience into our human world? (It need hardly be said that experiences that are not integrated into our emotional and imaginative lives later have strange and unpredictable repercussions.)[108]

Conclusion

The suggestions above, as well as some of those put forward in the individual chapters on the railroad, may be found on further consideration to be either impracticable or unfruitful. Some of them, however, undoubtedly point to important areas of research concerning the space program. For those who may enter upon this projected research, I should like to point out two cautions derived from our study of the railroad's impact on society as a historical analogy. The first is that the five generalizations listed earlier ought constantly to be borne in mind and, where possible, further tested and refined. For example, even our own research proposals concerning the space program listed above and in the other chapters do not sufficiently recognize the first generalization and its demand for multivariate analysis; or the fourth generalization and its insistence on a "stages" analysis. The second is that, as pointed out at the beginning, heed must repeatedly be paid in dealing with historical analogies to what I have called "historically conditioned awareness." For example, because people's religious beliefs have suffered severe shocks in the past, from Copernican or Darwinian theories, an immunity or insensitivity may now exist toward the impact of new discoveries.

If, however, careful studies are carried out in the light of the above admonitions, and then compared to the historical analogy of the railroad, we should have two rewards for our labor: (1) the beginnings of a more or less new branch of social study (or at least

---

[108] Cf. on this general subject the debate on "The Two Cultures," touched off by C. P. Snow's book of the same name.

a new approach to it), that is, the comparative history of the impact on society of social inventions; and (2) with this new discipline (indeed, it alone offers the requisite theory for approaching the empirical data), a more knowledgeable guide to the impact, specifically, of the space program, and with this latter knowledge, a better chance of guiding and controlling an important part of our future. Man's thrust into outer space is, ultimately, a return to himself.

2

A Technological Frontier: The Railway[1]

Thomas Parke Hughes

Wherever and whenever nature in her nonanimal manifestations frustrates man in the pursuit of his objectives, there exists a technological frontier. To penetrate the frontier man must develop techniques or a technology allowing him to adapt to, modify, or obliterate nature.[2] The most extreme result of technological frontier penetration is the creation of a man-made environment and the rendering of nature imperceptible. The phrases "man-the-maker" and "the second creation of the world" both imply the extreme effect. The front edge of technological advance delineates the border between the natural and the man-made world.

Here we are concerned with the penetration of a technological frontier through the use of the railway. The time is the nineteenth century and the place is the United States. The focus of attention will be America, but this emphasis should not be allowed to distort the fact that techniques applied in America had often been worked out elsewhere. The technological hinterland for the American rail-

---

[1] Like Bruce Mazlish I am suspicious of parallels between the past, present, and future, but I too see the possibility of moving up onto a level of abstraction where the terrain of the past is suggestive of the topography of the present and its future projection. The abstraction which seemed most suggestive to me of a relationship between the railway and the spaceway was that of a technological frontier. Approaching the railway with this concept I wrote the following chapter with the space program thrust as far back in my mind as possible. Undoubtedly, my layman's familiarity with the outline of the space program influenced the selection of my historical materials, but I hope my conscious effort to avoid the influence at least helped me shun the more superficial parallels. Only after the completion of my historical essay on the railway frontier did I attempt to raise credible questions — suggested study — about the space program (see my conclusion).

[2] Nature is used in this essay in the early modern sense of God-created, the physical environment which was not man-made or formed. "Natural laws" (man's intellectual abstractions) are not implied by the term as used here.

way frontier was not only the American technological base but the British and the European as well.

During the third decade of the nineteenth century, discussions of railways became common. The visionaries whose premature enthusiasm fed on grandiose concepts allowed by the dearth of experience had become less prominent; the prudent organizers and careful investigators were more influential. The railway to these men was a transportation system — often not involving a steam locomotive — which was economically competitive with canals and turnpikes in geographical areas where these were possible, and which was conceivable in areas where the canal was not possible.[3]

In the 1820's the emphasis was upon the *railway* element of the system, not the motive power; the railway was "a mere road, made of materials by which the resistance of friction is considerably reduced, whereby a propelling power is capable of more useful effect."[4] The creation of a railway system was thought of as involving the following complex of elements: surveys and leveling (the determination of topographical characteristics); route planning; superintending of construction; economically prudent right-of-way acquisition; design and construction of major facilities and minor (ditches, drains, culverts, fences, and gates); stone quarrying and cutting; manufacturing of rails, sleepers, and accessories; building of related facilities such as loading platforms and machines, fueling stations, and weighing stations; as well as the general problems of managing and financing operations.[5] To entrepreneurs and engineers the railway was far more than the "iron horse" with which it was later associated in the popular mind.

Frontier Problems and Related Techniques

One set of limitations imposed by nature and constituting a technological frontier arose from man's use of natural and artificial waterways as means of transportation. Since the beginning of history commerce had depended upon water for the transport of heavy and bulky goods over long distances. Thus, internal navigation was limited to those areas where water ran naturally or could be channeled artificially; and to those waterways where frosts were not

---

[3] George Armroyd, *A Connected View of the Whole Internal Navigation of the United States; Natural and Artificial, Present and Prospective* (Philadelphia, 1830), pp. 566–568.

[4] Report of Engineer C. Crozet on the navigation of the James River (Richmond, July 1, 1826), quoted in Armroyd, *op. cit.*, p. 569.

[5] Thomas Tredgold, *A Practical Treatise on Rail-Roads and Carriages* (2nd ed., London, 1835), pp. 139–140. Tredgold died in 1829.

severe and prolonged. Because of these natural conditions much of the inland region of the temperate zone constituted a technological frontier before the railway era. For example, a major disadvantage of the Erie Canal — extending as it did from the lake to the navigable Hudson at Albany — was the freeze-up for four or five months during a hard winter.[6] In an era when not only transportation but also mill power depended on water, the winter freeze greatly limited economic activity.

The use of a waterway or canal made man subject not only to the limitations imposed by freezing cold, but also to those imposed by drought and the unavailability of water. Though not subjected to the ice barrier, even the southern regions of the northern hemisphere knew this aspect of the technological frontier. Canals, for example, could not be run by locks over elevations unless feed water for the canal was available at the summit level. These circumstances would have precluded the participation of many areas in a modern transportation-based economy if the canal technique had not been superseded by the railway. The railway engineers transcended these natural limitations by substituting a "way" of rails for the contingent-upon-nature waterway.

In addition to cold and drought, another problem on the technological frontier was elevations.[7] One team of engineers surveying a railway route catalogued these as: undulations upon the surface of the ground; ravines and water courses (to be crossed); precipices and abrupt slopes; and the ascent of hills and mountains.[8] In the face of these natural barriers the railway had no clear advantage over the canal. As a matter of fact, the railway engineer had the precedent of the canal builder to spur him as he encountered the frustrating varieties of the earth's surface. If American and British canal engineers had not already begun the conquest of this limitation and made possible a carry-over of techniques from the canal frontier, railway advance would have been less rapid.

The more scientific engineers, in their analytical essays and their reports, abstracted the force of gravity as the common element in

---

[6] John Stevens, *Documents Tending to Prove the Superior Advantages of Rail-Ways and Steam-Carriages over Canal Navigation,* 1812. (Reprinted under joint auspices Harvard School of Business Administration and Railway and Locomotive Historical Society, Inc., Boston, 1936), p. 17.

[7] A critic of the Erie Canal proposal observed that the waterway had to expose itself to the colder northern route rather than a more southerly one because of the "elevations" to the south.

[8] William Howard and Others, *Report of the Engineers on the Reconnaissance and Surveys Made in Reference to the Baltimore and Ohio Rail Road* (Baltimore, 1828), pp. 168–174. Hereafter cited as the *B & O Report.*

elevations or earth surface barriers. To conceive of the problem as one of gravity allowed for a quantitative attack upon it, and also permitted a distinction to be made between the force of gravity and the force of friction,[9] a distinction and an application of science conducive to more effective engineering. The analytical analysis of the elevations barrier in terms of gravity allowed the engineer to apply mathematical techniques used by post-Newtonian scientists. Furthermore, the quantitative analysis allowed decisions to be made on an economic basis when alternative techniques were employable (money — like gravity force units — could be counted). Science rationalized the technique of the engineer as he attacked "elevations"; economics helped him to decide which technique to use.[10]

Various techniques were available for the attack upon the undulations of the earth's surface by the railway engineer. He could level these to be free of nature's force of gravity. This he did by cutting (excavating), filling (embanking), bridging, and tunneling to construct his railway. He could compromise with nature by cutting and filling only to the extent necessary to reduce gravity to manageable proportions, or he could compromise further by taking a circuitous route around the elevations (or depressions). He avoided the long span bridge and the tunnel because of the state of his art. The early American railway engineers tended to compromise with nature both because of the state of their technology, and because of economic and political factors.

The economic and political factors influencing technological decisions in less advanced America contrasted with those in Britain. The engineers surveying a route for the Baltimore & Ohio Railroad in 1828 acted on the assumption that the nation's political economy created an unlimited and immediate need for railways in America.[11] Others considering the untapped — almost unknown — resources of the interior also pressed for expedient engineering giving temporary

[9] "Man-derived" friction is caused by axles turning in inadequate bearings and wheel flanges running not-true on iron rails. One authority asserted that friction developed by steam carriages running on common roads would prohibit competition for railway transportation from that quarter. Peter Lecount, *A Practical Treatise on Railways Explaining their Construction and Management* (Edinburgh, 1839), p. 348.

[10] In the second edition of a book written in 1825, Englishman Nicholas Wood considered the following engineering problems in an analytical-quantitative way: experiments on the strength and deflection of cast and malleable iron-rails; friction of axles and wheels on rails; friction of ropes dragging carriages up inclined planes; experiments on various forms of power for effective performance [efficiency], etc. Nicholas Wood, *A Practical Treatise on Rail-roads* (2nd ed., London, 1831).

[11] *B & O Report*, p. 64.

works. It was assumed that techniques less conducive to early obsolescence could be used later when replacement was necessary. These sentiments were reinforced by the politically minded who considered railways and canals as ties binding the new nation. American railway engineers of the early period, stimulated by these beliefs and conditions, exhibited the same characteristics that fixed the label "go-ahead people" on American electrical engineers a half century later.[12] This "go-ahead" attitude of the railway pioneers stimulated compromises with nature's elevations. In his influential *Railway Economy: A Treatise on the New Art of Transport* (1850), Dionysius Lardner observed that neither the Germans nor the Americans attempted "the vast expenditure for earth work and costly works of art, such as viaducts, bridges, and tunnels, by which vallies are bestridden and mountains pierced to gain a straight and level line [as] in the English system." (p. 471)

The railway engineer who pushed ahead against natural limitations identified gravity as the essence of his "elevations" and then attacked the problem with his level roadbed; he also identified friction and then attacked this obstacle with the smooth rails he laid on his roadbed. Engineer C. Crozet, for example, making a survey in 1826 of transportation possibilities in the James River Valley, observed that "friction and inertia are, it is well known, the two great obstacles that engineers or mechanicians have to overcome in the application of power. If it were not for the constant resistance opposed by friction, bodies once set in motion on a plane would never stop." [13] Engineers believed that friction varied according to the substances placed in contact, was proportional to the weight moved, and remained virtually constant whatever the velocity. This concept also made it possible to demonstrate the advantage of a railway in quantitative terms that could be translated into dollars and used in making economic decisions.

Drawing upon the gifted British engineer, Thomas Tredgold (1788–1829), Crozet argued that the smooth rails made a horse more effective than would a macadamized road, and that at velocities over $4\frac{1}{2}$ miles per hour the railway made more effective use of motive power than the canal ("on a canal . . . the resistance in-

---

[12] *Electrical Review* (London), XVII (1885), 336, as cited in Thomas P. Hughes, "British Electrical Industry Lag: 1882–1888," *Technology and Culture*, III (1962), 37. In this article on the "British . . . Lag," another aspect of a technological frontier is analyzed.

[13] Crozet, quoted in Armroyd, *op. cit.*, p. 569. Evidence of this kind contradicts the stereotype that electrical engineering was the first engineering based on science.

creases as the square of the velocity of the boat. . . ."). Believing that the friction advantage of a railway in comparison to a canal began at velocities higher than the most effective velocity of a properly loaded horse, engineers seriously considered the locomotive as a substitute for the horse, and rails as a substitute for water in proportion to the importance attached to saving time. Thus, the engineers practicing in regions where production rates in the manufacturing sector were being maximized were better able to see the economic advantage of a steam locomotive on rails. Nevertheless the engineers' report for the B & O in 1828 about a railway in a non-manufacturing region left open the question of locomotion.

While the engineer practicing in America drew resourcefully upon British and continental science to analyze the friction problem, he tended to compromise more with nature in choosing his materials for the rail solution. Encountering rich resources in wood and stone, the American engineer planned to use these materials rather than to manufacture iron rails. Furthermore, the absence of manufacturing establishments and craftsmen caused him to think often in terms of horse and water power rather than of steam engines. The American engineer reverted in the use of materials and power to the long-passed era for his European predecessors of plentiful wood, and of wind, water, and animal power. From the European point of view, early American railroad engineering consisted of applied contemporary science on a medieval frontier.

Several of the earliest American railway projects manifested the tendency to compromise with nature. In 1812, John Stevens, who had already pioneered in utilizing steam for ship propulsion proposed a railway from Albany to Lake Erie with rails of timber held above the ground by wooden stanchions.[14] He intended to resort to iron plates to protect the wooden rail surfaces only if wear on the wood rail proved excessive. The Quincy railway, built to carry heavy granite three miles from the quarry to Boston harbor, had sleepers of granite upon which rested rails of pine, with a covering of oak and plates of iron. In 1828, the B & O engineers proposed, in their survey report, stone rails laid in gravel-filled trenches. Thin plate rails of wrought iron were to be fastened to the dressed upper surface by means of rivets, but the engineers recommended wooden sleepers and rails where construction was temporary.[15] These early

---

[14] Stevens, *op. cit.*

[15] Engineers were specific about the kinds of wood (locust, mulberry, cedar, white ash, or hard pine) to be used, as each possessed peculiar characteristics. *B & O Report*, pp. 60–62. Scientific engineer, Thomas Tredgold, wrote an essay on the strength of timber and other materials (1817). The use of wood does not imply "quaint" engineering.

wood and stone (sometimes iron-plated) American railways used materials found in abundance along the way.

The engineers also planned to build bridges of natural materials. The B & O engineers recommended wooden bridges of yellow pine resting on abutments and piers of stone. Not only was wood available in quantity along the westward routes, but such bridges could be erected expeditiously and repaired more easily than bridges of iron or arches of stone. Furthermore, wooden structures were thought less liable than stone structures to injury by freezing.[16]

Artfully negotiating with nature, the early engineer reduced the forces of gravity and friction to manageable proportions through the use of his nearly level roads and rails — but "manageable" sometimes implied the use of substantial power for locomotion. As he approached the Allegheny Mountains in his westward thrust from the major ocean ports and manufacturing districts, he found the power problem especially troublesome. The early engineer's utilization of power also manifested his flexibility, his use of science, and his ability to compromise with nature.

An example of his style and resourcefulness was the resort to the "inclined plane"[17] in combination with various power sources. The motive power for drawing wagons up a plane might be a stationary, condensing steam engine, but the engineer could remain closer to nature by using horses or water power to turn wheels or windlasses, and to draw in the attached cables. Circumstances were considered remarkably favorable if a head of water could be found near a precipitous elevation. Descending loads often counterbalanced those moving up. The nine-mile-long, inclined-plane railway completed in 1827 to bring out coal from the Lehigh coal veins to loading platforms at Mauch Chunk Creek,[18] and the Quincy granite railway (essentially an inclined plane), offered precedents in America for engineers contemplating inclined-plane sections in cross-country railways.

The two early inclined planes used horses and mules to furnish the necessary power.[19] Even if the engineers compromised with

[16] *B & O Report*, p. 69.
[17] The common use of the term "inclined plane" by the engineers shows that scientific principles underlay their techniques.
[18] Armroyd, *op. cit.*, p. 139. This short, railway inclined plane was part of a system of internal navigation which permitted the economical transportation of coal from the mines to the Philadelphia area. The system concept was used by James Brindley in the first major canal construction in Britain (c. 1760).
[19] On one occasion at the Mauch Chunk inclined plane, the carriages in which the animals descended were not drawn up with the empty coal wagons, and the animals refused to walk down. Laborers had to draw up the animal carriages.

nature by using animals rather than steam, they had available analytical treatises on animal power giving quantitative data derived from experience and deduced from scientific principles. In the 1820's and 1830's, horsepower — literally — was one of the major alternatives whether the railway was on an inclined plane or on a reasonably level roadbed; and there was a science of horsepower.

American engineers used data from Tredgold's *Practical Treatise on Rail-Roads and Carriages Shewing* [sic] *the Principles of Estimating their Strength, Proportions, Expense and Annual Produce. . . .* (London, 1835). Tredgold had conducted researches on horsepower that allowed him to tabulate comparative data and deduce scientific principles. Using simple formulas, for example, he calculated the continuous moving force to be obtained from a healthy horse working for six hours when different velocities were maintained for this period. His table ranged from 2 mph to $5\frac{1}{8}$ mph in increments of about $\frac{1}{2}$ mph. He also tabulated experimental data on the relationship between velocity and duration of labor. He found a six-hour work day for the horse moving 125 pounds at about 3 mph a reasonable utilization of horsepower.[20] He warned that indifference to the science of horsepower might extend to the railroads the barbarous practice of working horses to death.

The railway, in the first quarter of the nineteenth century, was indeed a promising technique for frontier penetration — for the taming of the natural frontier. John Stevens, in 1812, realized that "wind and tide, rough or smooth water, light or darkness, would have no influence whatever over steam-carriages moving on these ways."[21] Tredgold observed that "speed and certainty are of such primary importance in commerce," that a small increase of expense is not a material object, and that the "certainty of supply must tend much to diminish the fluctuation of prices, and remove those alternations of glut and scarcity which are perpetually occurring in the markets from contrary winds, frosts, floods, etc."[22]

The second creation — of a man-made world — had caught the imagination of the railway engineers and the society which supported them: horses that tired, land surfaces that undulated, weather conditions that frustrated, distances that consumed inordinate time, must all give way to a fixed, uniform environment conducive to an economic system involving co-ordination and prediction.

[20] Tredgold, *Practical Treatise*, pp. 70–72. The experiments cited ignored gravity.
[21] Stevens, *op. cit.*, p. 30.
[22] *Op. cit.*, p. 3.

The Railway Technique: An Acquired Engineering Characteristic

When the railway frontier was opened about 1830, the advances expressed tentative compromises with nature; the engineer often made his way up, over, and around. About fifty years later, the technique had been refined and the assault on nature was frontal: the engineer obliterated nature's earth surface undulations as he bored through mountains and spanned great depressions to maintain his unnatural level. At the opening of the century, man thought himself generally destined to move about on the surface of the earth;[23] by the last quarter, he had been conditioned to create his own level. Before considering the transferal by the engineer of the railway technique — acquired in solving the distance problem — to the space problem, we shall note the high degree of competence he had acquired after a half century of railway experience.

The great-span bridge and the long tunnel avoided by the engineers of the first quarter of the nineteenth century had become established techniques by the last quarter. The history of the tunnel extends back to surface mining and to early canals (the Languedoc Canal completed by 1681 had a summit-level tunnel 180 yards long), but the experience of the railway frontier contributed substantially to the engineering art of driving tunnels measured in miles rather than yards. Furthermore, after mid-century, tunnels were projected through rocks and soils that, earlier, would have been avoided. European (including British) engineers led in tunneling as they had in railway technology generally. Writing in 1866, an American called attention to the European experience available for American engineers and observed that railway tunnels were so common in Britain "that they have long ceased to attract the attention of travelers." [24]

In Europe, the great tunnel of the last quarter of the century was the Mt. Cenis, driven from newly French Savoy seven miles through the Alps to the soil of the Piedmontese province of newly united Italy. In America, its counterpart was the half-as-long Hoosac Tunnel, extending through the Appalachian barrier of northwestern Massachusetts. The Hoosac Tunnel, when completed in 1876, broke a new route from coastal Boston through to the burgeoning Midwest. It also made available the timber and water-power resources

[23] Earlier he had found a naturally level — gravity-free — surface on the ocean.
[24] John J. Piper, *Facts and Figures Concerning the Hoosac Tunnel* (Fitchburg, 1866), p. 17.

of northwestern Massachusetts. Before the required engineering techniques and engineering confidence developed, a more southerly compromising-with-nature route (the Western Railroad) had been the expedient.

Engineers faced new problems, and created the solutions, in driving the two tunnels. The problems facing the Mt. Cenis engineers — three young Italians, Sommeiller, Grattoni, and Grandis — are suggested by the skeptics who asked, "how were the workmen to breathe"; "what chasms, unfathomable abysses and resistless torrents might not be encountered"; was it certain that the two sections commenced from the opposite ends would not miss and pass each other in the middle of the mountain?" Geologists could not predict the temperature of the rock a mile below the surface, and they could not guarantee that overlying strata weakened by blasting might not collapse without warning. Similar doubts and unknowns also arose in connection with the Hoosac Tunnel.[25]

The techniques developed to solve the problems defy enumeration. Among the most famous was the use of the pneumatic drill frame. When both tunnels were begun, hand drills were used, as steam drills were impractical in an unnatural environment where the shortage of air was crucial. The drilling frame developed for the Mt. Cenis Tunnel (an adaptation was used later at Hoosac) had nine or ten pneumatically powered drills working simultaneously on the face of the heading. Tubes conveyed the air pressure miles from a water-powered air compressor (outside the tunnel) which was designed to work on the principle of a hydraulic ram. The drill holes were being charged with gunpowder at the time the tunnels were begun, but dynamite and nitroglycerine were being used by the time of completion. On other major railway tunnels constructed at the time, flooding posed another major problem which engineers attacked creatively (at Kilby Tunnel in Britain, water had to be removed at the rate of two hundred gallons per minute).[26]

The great railway tunnels and iron-steel bridges became schools of engineering as had the Erie Canal and the B & O Railroad earlier. Of American bridges, that of engineer James B. Eads at

[25] *Ibid.*, p. 18. In the great Alpine railway tunnels and the Hoosac Tunnel, the engineers and contractors broke down the *ancien régime* in underground work. Prior to the nineteenth-century railway tunnel experience, the tunnel constructor was in need of mining know-how; during the past hundred years the technological advance has been in tunneling and "it has been the tunnellers who have inspired the progress made in mining." Gösta Sandstrom, *The History of Tunnelling* (London: Barnè and Rockliff, 1963), p. 286.

[26] "Report of Chas. S. Storrow, on European Tunnels," *Hoosac Tunnel Reports* (Boston, 1863), p. 22.

St. Louis, across the Mississippi, provides a good example of engineering creativity and of learning through experience. St. Louis needed a railway bridge because bridges built to the north, where the river was less challenging, threatened to end the city's role as a trade center. The Eads bridge, completed two years before the Hoosac Tunnel, met the challenge, but only after Eads had developed new techniques that were necessitated by the uncompromising character of the structure. Among his major innovations were the pneumatic caisson which he introduced in the United States; steel members in the arches (he introduced the metal into American building techniques); and the erection of tubular arches without falsework by cantilevering them out from the piers.[27]

The great bridges and tunnels of the second half century obliterated the natural obstacle of elevations. The know-how gained — like that won earlier on the railway — became teachable to young engineers after it had been organized in a textbook way. In this manner, creative challenges became routines. As would be expected, the young engineers, taught the techniques in engineering schools, constituted a substantial addition to the engineering corps leveling the earth's surface. Less to be expected, though, was the transfer of these acquired characteristics to the creation of new surfaces above and below the natural surface. Before investigating this new frontier, however, we must show that the characteristics acquired by experience were being transmitted by formal education, with the rise of the engineering schools, and by the proliferation of texts and handbooks which rationalized and quantified the experience.[28]

Successive waves of railway engineers were trained in the engineering schools whose faculty organized and taught the information developed by the first wave. This was evident, in 1870, from the civil engineering courses offered in the four engineering schools

[27] Carl W. Condit, *American Building Art: The Nineteenth Century* (New York: Oxford University Press, 1960), pp. 185–190. The inadequacy of the cast-iron bridge to meet the needs of the British railway engineers led to the development of the wrought-iron bridge in the 1840's, and to its acceptance as the prevailing type until cheap steel became available. The condition conducive to invention and development is exemplified by the British railway engineer, Robert Stephenson, facing two large gaps at Menai and Conway on the Chester and Holyroad Railway, with no possibility of bridging them with cast iron — within a few years he had created, with the help of others, the tubular form of wrought-iron bridge. R. M. Sutherland, "The Introduction of Structural Wrought Iron," a paper read before the Newcomen Society in London, March 1964.
[28] These books were often written by Americans who acknowledged their debt to British engineering experience and French engineering sciences.

established before the war.[29] Rensselaer Polytechnic Institute had a complete curriculum in civil engineering in 1835, and by 1870 the course of study included geodesy (leveling, topographical surveying, road surveying), road engineering (common roads, railroads, canals, tunnels), topographical drawing (plans, profiles, and sections of railroad surveys), "constructions," and mechanics of solids (friction, strength of materials).

Sheffield Science School at Yale, the Lawrence Scientific School at Harvard, and the University of Michigan had announced curricula for engineers by 1847. By 1870, Sheffield taught laying out curves, and all field operations required in locating a line of road, establishing grade, and determining the amount of excavation and embankment necessary. In addition, the civil-engineering students at the school could study the science of construction, including strength of materials; the establishment of foundations, walls, and arches; the theory and detail of bridges, roof trusses, "etc.," in wood and iron; and the graphics of stone cutting.[30] Lawrence Scientific School had twenty engineering students among the thirty-five students listed. Wolcott Gibbs, Louis Agassiz, and Asa Gray were on the small faculty in 1870, but Henry Eustis had charge of engineering. The course in engineering included instruction in surveying, the nature and properties of building materials, and their application to the construction of railroads, canals, and bridges.[31] At the University of Michigan, in 1870, the engineering faculty taught the following subjects which were useful to the railway engineer: field engineering instruments, topographical and railroad surveying; theory of trussed girders, planes, and elevations of engineering constructions; stone cutting; theory of tubular, suspension, and arched bridges; and Gillespie's "Roads and Railroads." [32]

The reference to Gillespie in the Michigan curriculum is to a text which had nine editions by 1864.[33] The author, a professor of civil engineering at Union College, acknowledged his debt to British

[29] Rensselaer, Sheffield (Yale), Lawrence (Harvard), and University of Michigan, the oldest American engineering schools, were chosen for the limited sampling made here. The Morrill Act of 1862 provided land-grant funds for the establishment of a large number of engineering schools which, quite naturally, offered civil engineering training.

[30] *Sixth Annual Report of the Sheffield School of Yale College, 1870–71* (New Haven, 1871), p. 42.

[31] The Catalogue for Harvard University, 1870–1871 (Cambridge, 1870).

[32] The Catalogue for the University of Michigan, 1870–1871 (Ann Arbor, 1871).

[33] W. M. Gillespie, *A Manual of the Principles and Practice of Road-Making Comprising the Location, Construction, and Improvement of Roads (Common, Macadam, Paved, etc.) and Rail-Roads* (3rd ed., New York, 1850).

and French technological sources but wrote a text suited to the American challenge. Gillespie wrote more on roads than railroads, but pointed out that the latter demanded the highest technical competence — "the details of their [railways'] construction no longer belong to the community at large, but demand the highest professional skill of the Civil Engineer." [34] While road and railroad location followed the same principles, Gillespie observed, the railroad required much greater excavation, embankment, and bridges, in order that minor undulations of the country might be disregarded ("where a common road should go around a hill, a railroad should cut through it").[35]

Gillespie could serve as an introduction for the engineering student to the monographs of a more scientific or quantitatively precise character. Available to him then (around 1870) were books on calculating the cubic contents of excavation and embankment by diagrams; rules for the measurement of earthwork by prismoidal formulas; the theory of strains in girders and similar structures; the properties of limes, hydraulic cements, and mortars; and the properties of wrought iron and steel.[36]

Transfer of Technique from the Interurban to the Intra-urban Realm

By 1870, more than fifty thousand miles of railway had been opened in the United States, and there was to be an intense spurt of building in the 1870's and 1880's. Nevertheless, railway engineering was no longer front-edge engineering as the abundance of textbook literature on the subject shows. Basic problems had been solved and the solutions rationalized and made available for routine application. The railway technique was taken for granted as the means of overcoming natural obstacles and establishing routine and regular transportation over substantial distances (interurban transportation).

[34] *Ibid.*, p. 262. Gillespie's observation helps support my thesis that the substantial attack upon nature described here was made by the railway rather than road building. By mid-century, road building was a technology largely dependent in its development upon railroad engineering.

[35] *Ibid.*, p. 290.

[36] The availability of texts treating engineering problems in a scientific manner also demonstrates the error of maintaining that electrical engineering was the first engineering based on science. It is noteworthy, however, that scientifically inclined engineers recognized that there were many problems in engineering involving too many variables for science to solve them satisfactorily. See, George L. Vose, *Manual for Railroad Engineers* (Boston, 1873), p. vi. The empirical rule was the technique in these cases.

In the 1870's, man found a new challenge in the realm of intra-urban transportation (a space problem). Engineering characteristics developed over a half century on the interurban frontier would be transposed to the intra-urban realm. Furthermore, these characteristics, or acquired techniques, predisposed the engineer — and society — to choose the particular transportation solution worked out during the half century of railway building.

There were alternative solutions possible to urban congestion and the intra-urban transportation problem, and, if the railway technique had not been available, one of them might have been utilized.[37] For example, Peter Kropotkin, geographer and communist anarchist, in 1899 wrote about the possibility of utilizing electric power and communication to disperse the culture of cities and to limit their size. Among others, Ebenezer Howard also publicized the case for dispersal of population in his *Garden Cities of Tomorrow,* published in 1902, and much of Howard's proposal was already familiar.[38] The introduction of commercial electric power in the 1880's allowed either a two-dimensional (dispersal) or a three-dimensional (subway-elevated) solution.

The problem of urban congestion and intra-urban transportation intensified concomitantly with the growth of the railroads, and was, in part, an effect of the establishment of a railway network. Great concentrations of population in urban centers is not, however, a phenomenon peculiar to the industrial era and the advanced industrial nations, as is shown by the existence of great cities in the underdeveloped regions of Asia. Cities as religious, political, commercial, and financial centers flourished before the industrial era began. Industrial cities, however, such as Manchester and Pittsburgh, were spawned and cultivated by the rapid industrialization in the era of mobile and concentrated steam power. The industrial cities depended upon the railway for the import of raw materials and the export of finished products.

Within the industrial cities, the concentration of power in factories brought a density of working population (and related service and business personnel) hitherto unknown. As industrial, business, and commercial enterprises of the industrial complex took over residential property in the heart of the great cities, slums resulted,

[37] The alternative technology, however, involved more social planning than the railway solution, which is an example of nonsocially directed (autonomous) technological development.

[38] Lewis Mumford, *The City in History* (New York: Harcourt, Brace and World, 1961), pp. 514–524. Mumford discusses the alternatives to further urban population density made possible by the railway solution.

in which fantastically high tenant concentration permitted the economic use of costly urban property. A movement to the suburbs of those workers and lower-income business personnel who could afford the expense of commuting also took place. The swell of commuter traffic heightened as the rise of real wages allowed more persons to live in the suburbs and to work in the urban center. The railroads contributed to this trend by providing transportation for some suburbs, and by bringing commercial travelers, shoppers, sightseers, and pleasure seekers to the metropolis.[39]

The problem was not only one of commuting to, but also of moving about within the city. Movement from the main-line railway terminus to one's destination was particularly difficult in most cities. If the public authorities allowed the terminal in the city center, the city streets were intersected by surface railways, while if the authorities denied the railways access (as was the case in London after the Parliamentary Commissioners recommended against it in 1846),[40] the suburban railway commuter had a serious problem.

Horsedrawn urban coach lines with fixed routes were introduced in the first half century but, added to the private horsedrawn vehicles and the pedestrian traffic, the omnibuses found city thoroughfares prohibitively congested. Later, when horsedrawn streetcars were introduced, these used the surface thoroughfares. Writing in 1864 of New York City, A. P. Robinson, an engineer, observed that an extrapolation of the passenger trend in the city would mean that surface transportation would be impossible, "the streets would be absolutely blocked, and the time occupied by the trip would be a loss from the occupations of the city which would be unendurable." [41]

The introduction of electric streetcars in America in the late 1880's increased the velocity of street traffic and, therefore, the volume; but the fundamental problem remained unsolved: the limited surface area available for transportation facilities within cities whose population density (during the working hours) continued to rise with the introduction of multistory iron and steel-framed,

[39] C. M. Robbins and T. C. Barker discuss the problem of urban transportation in the nineteenth century in a *History of London Transport* (London, 1963), Vol. I.

[40] *Report of the Commissioners Appointed to Investigate the Various Projects for Establishing Railway Termini within or in the Immediate Vicinity of the Metropolis* (London: House of Commons, 1846).

[41] James Blaine Walker, *Fifty Years of Rapid Transit, 1864–1917* (New York: The Law Printing Company, 1918), p. 18.

elevator-served office buildings and crowded factories. People were working and living on many levels but commuting on the naturally limited surface. Here was a natural frontier to be transcended.

As it happened, the intra-urban transportation problem in America became most acute as main-line railway construction declined rapidly.[42] The two or three decades after 1880 when the three-dimensional, or elevated and subway transit, system had its substantial beginnings were the decades when the great cities in America grew most dramatically in population. The percentage of the population that was urban, in the United States, rose between 1880 and 1900 from 28 per cent to 40 per cent; New York City experienced the greatest decade of growth in its history between 1900 and 1910 (an increase of 1,330,000); Chicago experienced its greatest period of growth from 1890 to 1910 (599,000); and Boston also grew fastest during the same years (112,000). These cities pioneered in intra-urban transportation in America.[43] The technological (especially civil-engineering) focus was on the city.

The technology of subways and elevated railways resulted from the confluence of railway-building techniques and electrical power. London opened a steam locomotive subway in 1863 (the world's first) to join the peripheral main-line railway terminals, and New York had a steam locomotive elevated in the 1870's, but the use of steam underground and above streets near buildings was a noisome nuisance. The availability of reliable, direct-current motors of large horsepower initiated the period of rapid transit.

Most of the construction techniques and the mechanical elements of the elevated and the subway were, however, either borrowed directly from the interurban railway or were an extension of techniques learned on that technological frontier. The rapid-transit engineer drew upon the railway-bridge-building techniques to build his elevated way, and upon the railway tunnel experience to construct the subway. This carry-over of technique happened in several ways. The wave of engineers who flourished at the turn of the century when the subways and elevateds opened had been educated during the postwar years, when engineering curricula throughout the nation had been strongly influenced by the railway. The mono-

[42] Alfred D. Chandler, Jr., "The Beginnings of 'Big Business' in American Industry," *Business History Review*, 33 (1959), 1–31. Leland Jenks in "Railroads as an Economic Force in American Development," *Journal of Economic History*, IV (1944), 1–20, delimits the railway construction period as the 1830's through the 1890's.

[43] Statistics from the Thirteenth and Fourteenth *U.S. Census Reports*.

graphs and treatises used in solving subway and elevated construction problems were probably those of the railroad engineer.

Furthermore, engineers who had learned the profession in railway construction played leading roles in introducing rapid transit.[44] William J. McAlpine (1812–1890) served as the chief engineer of several railways, including the Erie, before planning an underground rapid transit system for New York City (he also projected underground, or second-level streets, at congested intersections). Theodore Cooper (1839–1919), after gaining experience as an engineer with the Troy and Greenfield Railroad, worked as a consultant for the New York and Boston Rapid Transit Company. Walter Katte (1830–1917) did engineering for several roads and was a division engineer of the Belvidere and Delaware Railroad before he went to New York City in the mid-1870's to become chief engineer of the New York Elevated Company. He built the first sections of the Third Avenue and Ninth Avenue elevated structures between 1877–1880. James Laurie (1811–1875), first head of the American Society of Civil Engineers and chief engineer of several railways, including the New Haven, Hartford, and Springfield, proposed an elevated railway for New York City in a paper, "The Relief of Broadway," read before the first regular meeting of the Society. Samuel Rea (1855–1929), who eventually became president of the Pennsylvania Railroad after beginning as a construction engineer, published a monograph in 1888 entitled *The Railways Terminating in London with a Description of the Terminal Stations and the Underground Railways*.[45] Rea, also chief engineer of the Baltimore Belt Railroad Company, brought the B & O into the heart of Baltimore by $1\frac{3}{4}$ miles of electrified tunnel in the 1890's.[46] Charles Sooysmith (1856–1916) spent two years with the Atchison, Topeka, and Santa Fe before becoming a consultant first on bridge foundations and then on the Underground Transit Railway of New York. Isaac C. Buckhout (1830–1874) was the superintendent of the New York and Harlem Railroad before planning an underground from Grand Central Depot to City Hall, and another in Brooklyn; after-

[44] While not directly to the point, it is relevant to note that Robert Stephenson, son of George, and one of the greats of railway engineering, proposed a subway or underground to carry main-line traffic from the peripheral terminals into the interdited heart of London. *Hearing of 1846 Commission* (London: House of Commons, 1846), pp. 17–28.

[45] Rea grasped the concept of a linked main-line interurban and underground intra-urban system.

[46] See *Engineering News*, 23 (1890), 605; and *Harper's Weekly*, 39 (1885), 821, for construction details.

wards, he became a member of the Committee on Rapid Transit. Harry B. Hanger, manager of construction for one of the country's outstanding contracting and engineering firms in the field of railway construction (Mason Company — later Mason and Hanger), worked on the Brooklyn subway, and the Independent Subway System of New York City.

Among the major subterranean transportation systems, constructed by these and other engineers who flourished near the turn of the century, were the access routes of the main-line railways into and through the heart of the crowded cities. The Baltimore Belt Railroad in 1895 initiated a trend which included an electrified route carrying the New York, New Haven, and Hartford Railroad into Grand Central Station (1906); a route for the Pennsylvania Railroad under the North and East Rivers; and the access routes to the Detroit and the Philadelphia terminals (completed by 1914).[47]

The first major subway in America opened in Boston in 1897, carrying passengers from Boylston Street, at the Public Garden, under the Common to Park and Tremont Streets. Contemporaries observed that, failing a subway, the Common would have had to have been cut open for street traffic; they compared the opening of the subway to "when a barrier is removed from the channel of a clogged-up river." [48]

While New York City pioneered in America with elevated railways, having constructed thirty-three miles by 1890, the Chicago elevated, extending for thirteen and a half miles in 1895, was judged a model at the time. The Chicago electric trains rode on a structure made entirely of basic open-hearth steel and incorporated the advanced techniques of the main-line railway bridge builders.

Therefore, by World War I, the railway technique, through a complex interconnected system of transportation, had been used to link the countryside, the cities, and the parts of the city to the vital center. The first wave of railway engineers had extended a technological frontier, not suspecting that the techniques learned there by experience would be formalized in the engineering curricula of the engineering colleges that proliferated after mid-century. Nor could they foresee that their techniques (which their main-line railways helped to create) would be applied to create a third dimension for the urban centers.

[47] T. Commerford Martin and S. L. Coles, *The Story of Electricity* (2 vols., New York: M. M. Moray, 1919–1922).

[48] The Boston Subway construction techniques are described in some detail in *Harper's Weekly*, 40 (1896), 1102–1103; 41 (1897), 934 ff. The techniques of open cut, cut-and-cover, and tunneling were all used on the first section.

Conclusion

The invention and development of the railway system was a century-long experience with psychological, institutional, ideological, sociological, and technological facets. As a result of the experience, Western man's Promethean confidence that he could recreate and order the world was heightened; his educational system felt the impact through curriculum changes and the demand for rational, teachable information in a broad new field; the security-engendering idea of co-ordinating disparate elements into a predictable action system was reinforced; and his mastery of new techniques utilized. All of these results were discernible in the nineteenth century.

Only later did the future-forming forces of the railway experience manifest themselves. The most sanguine railway enthusiast of 1840 did not foresee that the railway technique would help determine the subterranean and elevated solution for urban congestion. He did not envisage intersecting trunk lines that would generate a centripetal socioeconomic force intensifying the congestion that the subway was later designed to relieve. Only the professors at the "A and M's" spawned by the Morrill Act may have anticipated that the young railway engineers would find applications for railway technology when the constructional phase of the railway era had passed. (They knew that professionals could ingeniously ward off obsolescence in an environment that had a multitude of technological alternatives for solving a wide variety of pressing economic, social, and political problems by selecting alternatives with the bias of railway technology.)

The complex of persons, ideas, and institutions extending the railway system into the natural frontier established a momentum in history. Students of American history are familiar with Frederick Jackson Turner's frontier thesis. They know he believed that the western-frontier experience of the American nation lastingly shaped American character. Democracy and individualism were two of the most significant American character traits which Turner thought were reinforced by the frontier experience. Similarly, the aggressive-creative ordering of the material environment was a characteristic that was strengthened by the railway experience. Americans saw a wilderness become a building site. Under these circumstances the ancient deification of natural forces by the American psyche is unthinkable; a belief in the omniscience of "know-how" and the omnipotence of "can do" was inevitable.

Assuming that exploration of the space frontier will be a sustained national effort, we can draw hypothesis-generating analogies with the railway frontier. The space effort is analogous because technology is being used to penetrate nature's realm. Further, as pointed out elsewhere, the space effort now absorbs approximately the same percentage of the national product — and perhaps of the nation's attention — as the railway system during the constructional phase. Noteworthy is the analogous utilization of available engineering manpower by both. In addition, the railway and the space efforts have both occurred in eras when experiences tend to be institutionalized and reified. These are not eras when one can take it *and* leave it; the deeds of engineers and appliers of science survive — as machines, processes, and books.

By way of analogy, questions can be raised about the effects of the space effort. Because the space program has not moved beyond the opening "construction" phase, many of the questions (or suggestions for further research) arise from the availability of information about one major technological experience that has gone full term, and the unknown future of a developing one. This effort to extrapolate from a historical model is not based on the impression that "history repeats itself," but on the conviction that historical awareness sensitizes the observer to future probabilities, and, as pointed out in the introductory chapter, increases the probability by conditioning him to model his actions on the past. Some questions requiring further study are suggested here:

1. What style of engineering or engineering techniques will characterize the engineer or scientist who has learned in the "school" of NASA? What will be the kind of nonspace-effort technical problems to which he will be drawn after the constructional phase of the space effort has passed? How will this class of technical problem relate to major areas of human concern (for example, while it seems likely that the space-effort engineer will deal in sophisticated fashion with electronic control and communication phenomena, it seems improbable that he will be particularly well suited to provide canals and roads in underdeveloped areas)?

2. Will the acquired characteristics of the space-effort engineer after reification and institutionalization build up a powerful momentum analogous to that which carried the railroad techniques into urban transportation? Which engineering and science fields will languish because of the emphasis upon the space effort? Should NASA actively support the cultivation of some of these fields by other public and private agencies in order to ensure for the nation a balanced technology? Which of these fields needs the highest

support priority when evaluated in relationship to the NASA contribution (and noncontribution)?

3. Can the space effort be realistically envisaged as providing a complementary subsystem to our present overloaded transportation system (satellites have already provided an extension to our existing communications system)? Will the lifting of the transportation level (into space) give another congestion-relieving dimension analogous to that provided by the subway in lowering the level? What will be the economics of removing gravity and friction resistance to transportation far more effectively than was achieved by the railway roadbed and the smooth rail (the heavily capitalized launching facilities place the space vehicle in a gravity-free condition — when in orbit — comparable to the locomotive passing over a bridge or through a tunnel)?

4. Will the psychological impact of a technology, which envelops man in a self-created "skin" (space suit or capsule) with an internal environment elaborately controlled and regularized, and which allows him to penetrate an environment more hostile than that of the railway frontier, be analogous to the aggressive-creative impact of the railway triumph?

5. What natural resources (in space) can be exploited because of the availability of a transportation system (in the example of the railway, natural resource complexes beyond navigable waters became economical)? Can the temperature, pressure, gravity, and radiation of space be exploited (the chemical engineer, since the mid-nineteenth century, has increasingly used extreme-condition processes such as the Haber nitrogen fixation process)? Will the accessibility of space and extreme-condition chemistry have a fruitful confluence analogous to that of railway-tunnel and bridge technology, and electric power?

6. What influence will the contemporary style of American engineering and science have upon the American technology of space exploration (in the nineteenth century, the British contrasted American empirical technology with their own by labeling the American compromising-with-nature style as "go-ahead" technology)? Can an American style of space technology be contrasted with a Soviet style?

3

Railroads as an Analogy to the Space Effort: Some Economic Aspects

Robert William Fogel

Introduction[1]

Viewed as an investment, the central issue posed by the space effort is: Will the increase in national income made possible by the space program exceed the increase in income that would be obtained if the same resources were invested in other activities? To answer this question one has to know the increment to national revenue that will follow from space activities in all future years. Given this information, the present value of the stream of returns can be computed and compared with the cost of the program. If the present value of the returns exceeds the cost then it should, from the purely economic view, be undertaken. If not, it should, on the same view, be abandoned. Of course it is possible that political or military factors might dictate a course of action different from that implied by purely economic considerations. Political and military issues, however, are beyond the scope of this chapter.

If a study of the influence of railroads on American economic growth is to contribute to an evaluation of the investment decisions posed by the space effort, two conditions must be met. First, the analysis of railroads must focus on the incremental contribution of this innovation. Discussions merely of things the railroad did are of no use unless we know whether these services represent more or less than would have been contributed by alternative investments.

[1] This chapter draws on concepts and empirical findings developed in my book, *Railroads and American Economic Growth: Essays in Econometric History* (Baltimore: The Johns Hopkins Press, 1964). I am grateful to The Johns Hopkins Press for permitting me to make use of these findings. I have also benefited from comments by Raymond Bauer, Albert Fishlow, Edward Furash, Bruce Mazlish, and Peter Temin.

Second, one must be able to show how the effects of the space program may conform to, or deviate from, the types of effects attributable to railroads both in magnitude and in quality. While this chapter focuses primarily on the first of the aforementioned conditions, a concluding section will deal with some aspects of the second one.

From an economic point of view the central or primary feature of the railroad was its impact on the cost of inland transportation. Obviously, if the cost of rail service had exceeded the cost of equivalent service by alternative forms of transportation over all routes, and for all items, railroads would not have been built and all of their derived consequences would have been absent. The derived consequences of railroads can be divided into two categories. The "disembodied" consequences are those that followed from the saving in transportation costs per se and which would have been induced by any innovation that lowered transportation costs by approximately the amount attributable to railroads. The "embodied" consequences are those that are attributable to the specific form in which the railroads provided cheap transportation services.

The Effect of Railroads on the Availability of Resources

The change in the availability of resources is perhaps the most important of the disembodied effects of the railroad. Of course, all parts of the nation would have been physically penetrable even in the absence of railroads. However, without this innovation the cost of transportation to and from some areas might have been so great that, from an economic point of view, large sections of the land mass would have been nearly as isolated as the moon is from the earth. By reducing the cost of transportation, railroads increased the *economic accessibility* of various parts of the natural endowment of the United States. The question is: Which endowments were so affected and by how much?

Agricultural Land

Agricultural land was the most valuable of the natural resources of the United States in 1890. It was also more widely dispersed than coal, iron ore, oil, and other mineral deposits. Land in farms occupied one third of the national territory in 1890. No state devoted less than 1 per cent of the area within its borders to farming, and no single state contained more than 8 per cent of the nation's total farm acreage.[2] Under these circumstances one would expect to find that railroads were more essential in obtaining access to farmland than to other resources.

Even in the case of farmland, however, certain factors suggest that the incremental contribution of railroads was limited. One of these is the experience of the half century following the ratification of the Constitution. The occupation not only of the territory east of the Appalachians but also of that lying between the Appalachians and the Mississippi River was well under way before the coming of the railroad. As Kent T. Healy has pointed out, it was "water transportation, available first on natural waterways and later on canals" which made possible the "astonishing redistribution of population and economic activity" during the first four decades of the nineteenth century. By 1840, "before a single railroad had penetrated that area from the coast, some 40 per cent of the nation's people lived west of New York, Pennsylvania and the coastal states of the South."[3]

This western population was, of course, primarily engaged in agriculture. Two decades before the Civil War, Ohio was the chief wheat-producing state of the nation. It, together with Michigan, Indiana, and Illinois accounted for 30 per cent of the nation's total wheat crop. Moreover, with one out of every four of the bushels produced in these states being sold in the East and South, it is clear that commercial agriculture was well under way in the Old Northwest long before the era of substantial railroad construction.[4] The geographic locus of corn production provides an even more striking demonstration of the same point. Although they contained a bare total of 228 miles of disconnected railroad track, Michigan, Ohio, Kentucky, Tennessee, Indiana, Illinois, and Missouri produced 187,000,000 bushels of corn in 1840 — half the nation's total.[5] As for cotton, the westward movement of this culture was virtually com-

[2] U.S. Bureau of the Census, *Eleventh Census of the United States, 1890: Report on the Statistics of Agriculture in the United States*, V (1895), 74; U.S. Bureau of the Census, *Statistical Abstract of the United States, 1963* (84th ed., Washington, D. C., 1963), p. 173.

[3] Kent T. Healy, "American Transportation before the War between the States," *The Growth of the American Republic*, ed. Harold F. Williamson (New York: Prentice-Hall, 1944), p. 187.

[4] U.S. Bureau of the Census, *Sixth Census of the United States, 1840: Compendium* (1841), p. 358; Percy W. Bidwell and John I. Falconer, *History of Agriculture in the Northern United States, 1620–1860* (Washington, D. C.: Carnegie Institution of Washington, 1925), p. 329.

[5] *Ibid.*; L. Klein, "Railroads in the United States," *Journal of the Franklin Institute*, 30 (1840), 306.

pleted in 1850, by which time the geographic "limits of the cotton belt were practically the same as they are at present [1900]." Yet, during the entire ante bellum period the transportation of cotton "was conducted almost exclusively by means of water." As late as 1860, about 90 per cent of all cotton shipped to New Orleans arrived by boat or barge.[6]

The fact that the initial occupation of the trans-Appalachian lands was based almost exclusively on waterways is suggestive; but it by no means proves that waterways could have sustained later developments. The acreage of agricultural land in the North and South Central states underwent a fourfold expansion between 1850 and 1890.[7] It is therefore necessary to devise a method of determining how much of the land settled after the advent of railroads would have been settled in their absence.

Without railroads the high cost of wagon transportation would have limited commercial agricultural production to areas of land lying within some unknown distance of navigable waterways. It is possible to use the theory of rent to establish these boundaries of feasible commercial agriculture in a nonrail society. Rent is a measure of the amount by which the return to labor and capital on a given portion of land exceeds the return the same factors could earn if they were employed at the intensive or extensive margins. Therefore, any plot of land capable of commanding a rent will be kept in productive activity. It follows that, even in the face of increased transportation costs, a given area of farmland will remain in use as long as the increased costs incurred during a given time period do not exceed the original rental value of that land.

Given information on the quantity of goods shipped between farms and their markets, the distances from farms to rail and water shipping points, the distances from such shipping points to markets, and the wagon, rail, and water rates, it is possible to compute the additional transportation costs that would have been incurred if farmers attempted to duplicate their actual shipping pattern without railroads. In such a situation shipping costs would have risen not because boat rates exceeded rail rates but because it usually required more wagon transportation to reach a boat than a rail shipping point. In other words, farms immediately adjacent to navigable waterways would have been least affected by the absence

[6] "The Cotton Trade of the United States and the World's Cotton Supply and Trade," U.S. Bureau of Statistics (Treasury Department), *Monthly Summary of Commerce and Finance* (March 1890), pp. 2551, 2563, 2564.
[7] *Eleventh Census, Agriculture*, pp. 92, 100.

of rail service. The further a farm was from a navigable waterway the greater the amount of wagon transportation it would have required. At some distance from waterways the additional wagon haul would have increased the cost of shipping by an amount exactly equal to the original rental value of the land. Such a farm would represent a point on the boundary of feasible commercial agriculture. Consequently, the full boundary can be established by finding all those points from which the increased cost of shipping by alternative means the quantities that were actually carried by railroads is equal to the original rental value of the land.

This approach, it should be noted, leads to an overstatement of the amount of land falling beyond the "true" feasible boundary. A computation based on the actual mix of products shipped does not allow for adjustments to a nonrail technology. In the absence of railroads the mix of agricultural products would have changed in response to the altered structure of transportation rates. Such a response would have lowered shipping costs and hence extended the boundary. The computation also ignores the consequence on the level of prices of a cessation in agricultural production in areas beyond the feasible region. Given the relative inelasticity of the demand for agricultural products, the prices of such commodities would have risen in the absence of railroads. The rise in prices would have led to a more intensive exploitation of agriculture within the feasible region, thus raising land values. The rise in land values would have increased the burden of additional transportation costs that could have been borne and shifted the boundary of feasible commercial agriculture further away from water shipping points.[8]

The method outlined above has been used to establish the boundary of feasible commercial agriculture for 1890. In this year the relative advantage of railroads over alternative forms of transportation was probably greatest. During the half century that preceded it, increases in productivity reduced the cost of railroad transportation more rapidly than that of boats and wagons. The selected year also precedes the emergence of motor vehicles as an effective alternative. It thus appears likely that the incremental contribution of railroads to the accessibility of agricultural lands was at or near its apex in 1890.

Analysis of the relevant data indicates that in the absence of

---

[8] For a more detailed discussion of the theoretical issues involved in, and of the computational methods employed for, the establishment of the feasible boundary see Fogel, *op. cit.*, Chapter 3.

railroads the boundary of feasible commercial agriculture would have been located at an average of about forty "airline" miles from navigable waterways. Forty-mile boundaries drawn around all natural waterways used for navigation in 1890, as well as all canals built prior to that date, brought less than half of the land mass of the United States into the feasible region.[9] However, as Table 3.1

TABLE 3.1

FARMLAND LYING BEYOND THE FEASIBLE REGION OF AGRICULTURE, 1890 *
(thousands of dollars)

Region†	1 Value of Farmland 1890	2 Value of Farmland Beyond Feasible Region 1890	3 Col. 2 as Percentage of Col. 1	4 Value of Land Beyond Feasible Region as Percentage of Value of All Agricultural Land
North Atlantic	1,092,281	5,637	0.5	0.07
South Atlantic	557,399	117,866	21.1	1.45
North Central	4,931,607	1,441,952	29.2	17.75
South Central	738,333	158,866	21.5	1.96
Mountain	129,655	123,016	94.9	1.51
Pacific	671,297	95,200	14.2	1.17
United States	8,120,572	1,942,537	23.9	23.91

* For sources and method of computation see Fogel, *op. cit.*, Chapter 3, especially Table 3.7.
† The states included in each of the regions are *North Atlantic:* Maine, New Hampshire, Vermont, Massachusetts, Rhode Island, Connecticut, New York, New Jersey, Pennsylvania; *South Atlantic:* Delaware, Maryland, District of Columbia, Virginia, West Virginia, North Carolina, South Carolina, Georgia, Florida; *North Central:* Ohio, Indiana, Illinois, Michigan, Wisconsin, Minnesota, Iowa, Missouri, North Dakota, South Dakota, Nebraska, Kansas; *South Central:* Kentucky, Tennessee, Alabama, Mississippi, Louisiana, Texas, Arkansas; *Mountain:* Montana, Wyoming, Colorado, New Mexico, Arizona, Utah, Nevada, Idaho; *Pacific:* Washington, Oregon, California.

indicates, the feasible region includes 76 per cent of all agricultural land by value. The discrepancy is explained by the fact that over one third of the United States was located between the hundredth meridian and the Sierra Nevada Mountains. While this vast area fell almost entirely beyond the feasible range, it was of extremely limited usefulness for agricultural purposes. By value the area represented only 2 per cent of all agricultural land in use in 1890.

[9] A map of the area falling within the feasible region is given in *ibid.*, Figure 3.4.

Table 3.1 also shows that barely one quarter of 1 per cent of the lost agricultural land was located in the North Atlantic region, while only 6 per cent was in the Mountain states. About 75 per cent of the loss was concentrated in the North Central region. Indeed more than half of all the land lost by value was located in just four states: Illinois, Iowa, Nebraska, and Kansas. This finding does not support the frequently met contention that railroads were essential to the commercial exploitation of the prairies. The prairies were occupied at a time when the railroad had achieved clear technological superiority over canals. Consequently, the movement for canals that played so important a part in the development of the Eastern states was aborted in the prairies. The fact that major loss of land was concentrated in a compact area suggests an entirely different conclusion: a relatively small extension of the canal system would have brought into the feasible region most of the productive land that Table 3.1 puts outside of it.

Indeed, it would have been possible to build in the North and South Central states a system of thirty-seven canals and feeders totaling five thousand miles. These canals would have reduced the loss of agricultural land occasioned by the absence of railroads to just 7 per cent of the national total. The system would have been technologically feasible and economically profitable. Built across the flatlands of the middle west, the average rise and fall per mile of the proposed waterways would have been less than that which prevailed on all canals successful enough to survive railroad competition through 1890. The water supply would have been more than ample. And in the absence of railroads, the social rate of return on the cost of constructing the system would have exceeded 45 per cent per annum.[10]

The loss in agricultural land could have been further reduced by improvements in common roads. According to estimates published by the Office of Public Roads in 1895, regrading and resurfacing of public roads could have reduced the cost of wagon transportation by 50 per cent. This implies that the boundary of feasible commercial agriculture would have fallen eighty miles from navigable waterways. Together with the proposed canals, road improvements would thus have reduced the loss of agricultural land to a mere 4 per cent of the amount actually used in 1890.[11]

It thus appears that while railroads did increase the availability

[10] *Ibid.*, pp. 92–100, 218.
[11] *Ibid.*, p. 110.

of agricultural land, their incremental contribution was — even at the apex of railroad influence — quite small.

Iron Ore and Coal

Unlike agricultural land, which in 1890 occupied nearly one million square miles of territory in over two thousand counties in every state of the nation, the mining of iron ore was highly localized. The volume on mineral industries prepared for the Eleventh Census reported that the[12]

ranges embraced in the Lake Superior region are none of them of great extent geographically, and if a circle was struck from a center in Lake Superior with a radius of 135 miles, all of the present iron-ore producing territory of that region would be embraced within one-half of the circle, and most of the deposits would be near the periphery. The output of this section in 1889 was 7,519,614 long tons. A parallelogram 60 miles in length and 20 miles in width would embrace all of the mines now producing in the Lake Champlain district of northern New York, whose output in 1889, aggregated 779,850 long tons. A circle of 50 miles radius, embracing portions of eastern Alabama and western Georgia, included mines which in 1889 produced 1,545,066 long tons. A single locality, Cornwall, in Lebanon county, Pennsylvania, contributed 769,020 long tons in 1889. . . . In the areas named, which are only occupied to a limited extent by iron-ore mines, there were produced in 1889 a total of 10,613,550 long tons, or 73.11 per cent of the entire output of iron-ore for the United States.

Each of these major iron-producing areas would have been economically accessible in the absence of railroads. Cornwall was within five miles of the Union Canal and could have had a direct connection with that waterway.[13] The Coosa River, which was actually made navigable as far east as Rome, Georgia, flowed past the main iron-ore deposits of eastern Alabama and western Georgia. The red hematite deposits of central Alabama could have been reached by a relatively short canal built northward from the Alabama River to Shades Creek on the Cahawba River.[14]

As for the Lake Superior ores, their exploitation, beginning in

---

[12] U.S. Bureau of the Census, *Eleventh Census of the United States, 1890: Report on Mineral Industries*, p. 9.
[13] *Cram's Standard American Railway System Atlas, 1892* (New York, 1892), p. 51.
[14] *Eleventh Census, Mineral Industries*, p. 3; U.S. Bureau of Corporations, *Report of the Commissioner of Corporations on Transportation by Water in the United States* (3 vols., Washington, 1909–1913), Vol. I, pp. 85–86; Thomas Dunlap, ed., *Wiley's Iron Trade Manual* (New York, 1874), pp. 446–447; U.S. Geological Survey, *Water Supply Paper, No. 1384*.

1854, preceded the construction of railroads in the northern parts of Michigan and Wisconsin. Five years later, although the region was still without railroad service, its mines accounted for over 4 per cent of the national production of iron ore.[15] Here, too, the water supply and terrain would have permitted the construction of canals that directly linked the iron deposits of Michigan, Wisconsin, and Minnesota with the Great Lakes. A canal built along the Menominee River to Florence, Wisconsin, would have traversed the iron-mining district of the Menominee Range. A canal built along the Escanaba River to Ishpeming would have pierced the center of the mining area in the Marquette range. Canals could also have been built along the Montreal River and to Vermilion Lake in Minnesota.[16]

Many of the smaller deposits of iron ore were also well located with respect to water transportation. The main carbonate deposits of southern Ohio and northern Kentucky were located at the Ohio River; Tennesee's red hematite fields straddled the Tennessee River; and Pennsylvania had important iron ranges located along the Allegheny River, the Susquehanna River, and various of the state's canals.[17] It is not necessary, however, to consider all of the smaller deposits in detail: if one subtracts from the total domestic production of ore in 1890 that part required for railroad iron, the residual is about equal to the ore production of the major fields singled out by the Eleventh Census.[18] Undoubtedly, in the absence of railroads, the iron consumption of boats and other nonrail forms of transportation would have risen. But it is unlikely that such alternative

---

[15] Dunlap, *op. cit.*, p. 462; U.S. Bureau of the Census, *Eighth Census of the United States, 1860: Manufactures of the United States in 1860*, p. clxxvii; Frederic L. Paxson, "The Railroads of the 'Old Northwest' before the Civil War," *Transactions of the Wisconsin Academy of Science, Arts and Letters*, XVII (1914), 266.

[16] *Eleventh Census, Mineral Industries*, p. 3; *Cram's Standard American Railway Atlas*, pp. 98, 154, 160; *Water Supply Paper, No. 1387*.

[17] *Eleventh Census, Mineral Industries*, p. 3.

[18] Data in the *Eleventh Census* indicate that it required an average of about two tons of iron ore to produce one ton of rolled steel and iron. In the census year of 1890, some 2,092,000 tons of rails were produced. Thus, rails alone consumed about 4,200,000 tons of ore. Railroads required an additional one ton of iron for every two tons used as rails. At this rate, the total ore consumption of railroads in 1890 was 6,300,000 tons. However, railroads generated scrap which was the equivalent of 2,300,000 tons of ore. Hence the net ore consumption of railroads was about 4,000,000 tons. Subtracting the last figure from domestic ore production leaves a residual of 10,400,000 tons, an amount which is slightly less than the output of the four main centers of production referred to by the Census. U.S. Census Bureau, *Eleventh Census of the United States, 1890: Report on Manufacturing Industries*, Part III, pp. 402, 410, 412, 417; *Eleventh Census, Mineral Industries*, p. 9; Fogel, *op. cit.*, Table 4.8.

demand would have accounted for more than a fraction of the ore consumed in the production of railroad iron.

The production of coal, like that of iron ore, was highly localized. Nine states accounted for about 90 per cent of all coal shipped from mines in 1890. And within these states production was further localized in a relative handful of counties. Forty-six counties shipped 76,000,000 tons — 75 per cent of all of the coal sent from mines in the nine states. Moreover, all of these counties were traversed by navigable rivers, canals actually constructed, or the proposed canals discussed above. In this case too, many of the smaller deposits were well located with respect to water transportation.[19] Hence, it seems likely that a nonrail society could have had low-cost access to all of the coal it required.

The Position of Railroads in the Market for Manufactured Products

While the railroad was the chief vehicle by which late nineteenth-century society actually achieved access to natural resources, other mediums could have fulfilled essentially the same function. However, the low-cost services of these alternative forms of transportation were embodied in forms of equipment and structures different from those characteristic of railroads. Hence, it is still possible that railroads profoundly affected the course of economic growth because of the specific inputs, particularly of manufactured goods, required to produce railroad services.

Railroad input requirements could have affected the productivity of manufacturing in two ways. First, the railroad's incremental consumption of the output of various industries could have been so large that it moved these industries to a level of production permitting significant economies of scale. Second, the railroads could, uniquely, have induced changes in the technology of manufacturing processes that affected the production not only of railroad goods but also of goods consumed by other sectors of the economy. This section examines the first line of possible influence. The next section deals with the nexus between railroads and technological innovation.

The Ante Bellum Period

Iron is frequently cited as the classic ante bellum case of an industry raised to modern status by railroad purchases. Hofstadter,

[19] *Eleventh Census, Mineral Industries*, pp. 347, 348, 355–417; Fogel, *op. cit.*, Figure A-1.

Miller, and Aaron, for example, report that the railroad was "by far the biggest user of iron in the 1850's," and that by 1860 "more than half the iron produced annually in the United States went into rails" and associated items.[20] Such statements, however, are not based on systematic measurements but on questionable inferences derived from isolated scraps of data. Casual procedures have led to the use of an index that grossly exaggerates the rail share, to the neglect of the rerolling process, and to a failure to consider the significance of the scrapping process.

Since the iron industry did not produce one homogeneous product, and since railroads consumed various types of iron, the usual problem of how to aggregate these products arises. A desirable procedure would be to aggregate by prices and to use

$$I_1 = \frac{\text{value of domestically-produced railroad iron}}{\text{value of all final products of the iron industry}}$$

as an index of the proportion of the output of the American iron industry consumed by railroads. The numerator of I_1 is defined to exclude railroad iron purchased from abroad. The denominator is defined to exclude double counting. Unfortunately the available data are not complete enough to permit the construction of this index. Even for the years following the Civil War, the breakdown of production by type of product is not detailed enough, the prices of many individual products are not available.

As a consequence, writers dealing with the impact of railroads on the iron industry have resorted to indices based on the tonnage of iron production and consumption. A frequently used measure is

$$I_2 = \frac{\text{tons of iron used in the construction \& maintenance of railroads}}{\text{tons of pig iron produced}}$$

The implicit assumption made in using I_2 is that $I_1 = I_2$. This assumption would be true if (1), all railroad iron was purchased from domestic producers; (2), the amount of pig iron required to produce a ton of more highly manufactured iron was the same for all products; (3), only pig iron was used to produce more highly manufactured iron; and (4), the values of the final products of the iron industry were proportionate to the amounts of pig iron used in their production.

In actual fact all of these conditions were violated in such a way that I_2 is greater than I_1 by a substantial amount. A large part

[20] Richard Hofstadter, William Miller, and Daniel Aaron, *The American Republic* (2 vols.; Englewood Cliffs, N.J.: Prentice-Hall, 1959), Vol. I, p. 557.

of the railroad iron consumed through the 1870's was purchased from abroad. During the 1850's foreign rails represented nearly two thirds of all rail purchases; in 1871 they still accounted for 37 per cent of purchases. The pig-iron requirement of a ton of rolled iron differed from that of cast iron and hammered iron. Pig iron was not the only form of crude iron used in the production of final products; scrap iron became an increasingly important part of total crude-iron consumption in the years following 1850. Finally, the value of all final products was not directly proportional to the amount of pig iron required to produce them; the ratios of the price of a ton of steel and the price of a ton of hammered bar to the amount of pig used in their production exceeded the ratio of the price of a ton of rolled bar to its pig-iron content. Consequently, the use of I_2 as a measure of the share of output of the domestic iron industry consumed by railroads contains a considerable upward bias.

The neglect of the replacement process has also served to exaggerate the importance of railroads in the market for iron. Replacements became a major factor in rail consumption very early in the history of railroads. In fifteen of the thirty years following 1839, replacements represented more than 40 per cent of total rail requirements; in five of these years replacements accounted for two thirds of requirements. However, most replaced rails were scrapped. The availability of such scrap metal spurred the development of the rerolling of old rails. As early as 1849, one fourth of all domestically produced rails were rerolled from discarded ones. By 1860, rerolling accounted for nearly 60 per cent of domestic rail production.[21] Thus, although replacements rapidly became a substantial part of total rail consumption, replacement demand had little effect on the growth of blast furnaces. Replacements generated their own supply of crude iron. And scrapped rails that were not rerolled supplanted pig iron as an input in the production of other products.

When corrections are made for the bias of traditional indices, and when account is taken of the replacement process, it turns out that the net addition to pig-iron production attributable to rails during 1840–1860 amounted to less than 5 per cent of the output of blast furnaces. The significance of the railroads appears somewhat greater if account is taken of all forms of railroad consumption of iron from all sectors of the iron industry. On this basis railroads accounted for an average of 17 per cent of total iron production during the two decades in question. While it is true that the railroad

[21] Fogel, *op. cit.*, p. 194.

share rose to 25 per cent in the final six years of the period, more germane is the fact that during the quinquennium ending in 1849 railroad consumption of domestic crude iron was just 10 per cent of the total. Even if there had been no production of rails or railroad equipment whatsoever, the domestic crude iron consumed by the iron industry would have reached an average of 700,000 tons in the second quinquennium. The rise over the previous quinquennium would still have been 338,000 tons — an increase of 94 per cent, as opposed to the 99 per cent rise that took place with the railroads. Clearly the new high level of production attained by the iron industry during 1845–1849 did not depend on the railroad market.[22]

Furthermore, the demand for iron by industries other than railroads was more than adequate to permit a firm size capable of realizing the economies of scale in the production of nonrailroad iron that were actually achieved prior to 1860. The average capacity of blast furnaces prior to the Civil War was 4,800 tons; and the capacity of the largest furnace was under 10,000 tons. Similarly, except for rolling mills engaged exclusively in the rolling of rails, no firm appears to have produced more than 13,000 tons; but even if one assumes that the optimum firm capacity was 20,000 tons (the product of the largest rail firm in 1856 was 18,600 tons), the nonrail production of rolled iron at the time would have permitted seventeen firms of such size.[23]

Railroads occupied an even more modest position in the markets of other ante bellum manufacturing industries. In the case of lumber, railroads purchased a minuscule share of total output — this despite the large quantities of wood consumed as fuel and in the construction of track. The paradox is partly explained by the fact that the wood burned in the fireboxes of railroad engines was not lumber. A similar consideration is involved in connection with the railroads' consumption of cross ties. Throughout the nineteenth century railroad men believed that ties hewed by axe would resist decay better than sawed ties. Consequently, lumber mills were supplying ties amounting to only 450,000,000 feet B.M. during the last two decades of the ante bellum era. This was less than one half of 1 per cent of all lumber production. When the lumber required for car construction is included, the figure rises by half a point, to 0.96 per cent. The modest position of railroads in the

[22] *Ibid.*, pp. 130–135, 199.
[23] American Iron and Steel Association, *Bulletin of the American Iron Association* (Philadelphia, 1856–1858), pp. 58–63, 79, 103, 107, 155, 171, 173.

market for lumber products emphasizes the scale of lumber consumption by other sectors of the economy.[24]

The share of the output of the transportation equipment industry purchased by railroads is also surprising. From 1850 through 1860 some 26,300 miles of new track were laid. During the same time, about 3,800 locomotives, 6,400 passenger and baggage cars, and 88,600 freight cars were constructed. Yet, value added in the construction of railroad equipment in 1859 was only $12,000,000, or 25.4 per cent of value added by all transportation equipment. The output of vehicles drawn by animals was still almost twice as great as the output of equipment for the celebrated iron horse.[25]

As for other types of machinery, railroads directly consumed less than 1 per cent. Again, the situation does not change appreciably if indirect purchases at more remote levels of production are considered. When the share of machinery consumed by the lumber, iron, and machine industries that can be attributed to the railroad is added to that of transportation equipment, the railroad still accounts for only about 6 per cent of machine production in 1859.[26]

The transportation-equipment, rolling-mill, blast-furnace, lumber, and machinery industries were the main suppliers of the manufactured goods purchased by railroads. Using value added as a measure, railroads purchased slightly less than 11 per cent of the combined output of the group in 1859. Since these industries accounted for 26 per cent of all manufacturing in that year, railroad purchases from them amounted to a mere 2.8 per cent of the total output of the manufacturing sector. Railroad purchases from all the other manufacturing industries raised the last figure to just 3.9 per cent.[27] This amount hardly seems large enough to attribute the rapid growth of manufacturing during the last two ante bellum decades to the requirements for building and maintaining railway systems.

The Post-Civil War Era

Since the real capital stock of railroads increased at approximately the same rate as the real output of manufacturing, a detailed survey of changes in the position of the railroad in the market of most industries between 1859 and 1899 need not be undertaken in this

[24] Fogel, op. cit., p. 137.
[25] Ibid., p. 139.
[26] Ibid., p. 140.
[27] Ibid., pp. 145, 146.

paper.[28] The iron and steel industry, however, requires further consideration. It is frequently said that the introduction of the Bessemer process radically reduced the cost of producing steel and

TABLE 3.2

THE PRODUCTION AND CONSUMPTION OF STEEL, 1871–1890
(thousands of net tons)

Year	1 Production of Crude Steel	2 Steel Consumed in Rails	3 Steel Consumed in All Other Production	4 Col. 2 as Percentage of Col. 1
1871	82	44	38	52
1872	160	108	52	67
1873	223	147	76	66
1874	242	166	76	69
1875	437	332	105	76
1876	597	471	126	79
1877	638	494	144	77
1878	820	640	180	78
1879	1,048	792	256	76
1880	1,397	1,107	290	79
1881	1,779	1,549	230	87
1882	1,945	1,670	275	86
1883	1,874	1,481	393	79
1884	1,737	1,279	458	74
1885	1,917	1,234	683	64
1886	2,870	2,022	848	70
1887	3,740	2,713	1,027	73
1888	3,247	1,781	1,466	55
1889	3,792	1,937	1,855	51
1890	4,790	2,396	2,394	50

Column 1: Figures taken from American Iron and Steel Association, *Annual Report of the Secretary, 1889,* p. 63; *Annual Report, 1890,* p. 48.
Column 2: Total production of steel rails multiplied by 1.143. A.I.S.A. *Annual Report, 1886,* p. 43; *Annual Report, 1888,* pp. 36, 61; *Annual Report, 1896,* p. 68.
Column 3: Column 1 minus Column 2.

ushered that industry into a new era. As measured by value added, the production of basic steel products rose from 4 per cent of the

[28] Albert Fishlow, "Productivity and Technological Change in the Railroad Sector, 1840–1910," unpublished paper presented to the Conference on Research in Income and Wealth, Chapel Hill (September 1963), Table 6; Robert E. Gallman, "Commodity Output, 1839–1899," Conference on Research in Income and Wealth, published in *Trends in the American Economy in the Nineteenth Century,* Vol. 24 of *Studies in Income and Wealth* (Princeton: Princeton University Press, 1960), p. 43.

output of the iron industry in 1859 to 23 per cent in 1880.[29] Moreover, the consumption of steel was dominated by rails. Table 3.2 shows that in 1871 some 52 per cent of all steel ingots were consumed in the production of rails. The rail share rose steadily from that date until 1881, when it stood at 87 per cent, after which it declined to 50 per cent in 1890.

The fact that the rail share of steel production fluctuated between 50 and 87 per cent for a period of twenty years suggests that the market for rails was indispensable to the emergence of a modern steel industry in the United States. This opinion also seems to have been nourished by the inverse relationship between the average output of Bessemer mills and the prices of the products of these mills (see Table 3.3), a relationship that suggests economies of scale.

TABLE 3.3

THE AVERAGE PRODUCT OF BESSEMER STEEL MILLS AND THE AVERAGE PRICE OF STEEL RAILS

Year	1 Average Product of Bessemer Mills (in thousands of net tons)	2 Average Price of Bessemer Steel Rails in Dollars of 1890 (dollars per net ton)
1870	21	73
1880	90	62
1890	99	36

Column 1: The 1870 entry refers to a calendar year while the entries for 1880 and 1890 represent production in census years. In computing average product, furnaces using the Clapp-Griffith and Robert-Bessemer processes were excluded. Thomas Dunlap, ed., *Wiley's Iron Trade Manual* (New York, 1874), p. 187; A.I.S.A., *Annual Report, 1896*, p. 69; *Eleventh Census, Manufacturing*, Part III, pp. 411–412.

Column 2: Calendar year prices were deflated by the Warren-Pearson wholesale price index. The current dollar prices for the three years were $120, $76, and $36 per net ton. A.I.S.A., *Annual Report, 1896*, p. 84; U.S. Bureau of the Census, *Historical Statistics of the United States, Colonial Times to 1957* (Washington, D. C.; 1960), p. 115.

There is, however, another way of looking at the data contained in Table 3.2. Stress on the share of total output consumed by rails in any given year beclouds the extremely rapid rate at which non-

[29] *Eighth Census, Manufacturing*, pp. clxxviii, clxxx, clxxxiii, clxxxv, clxciv; U.S. Bureau of the Census, *Tenth Census of the United States, 1880: Report on the Manufactures of the United States*, II (1883), 748, 749, 753, 755, 757, 758, 759, 760. The term "iron industry" as used here includes forges, bloomeries, blast furnaces, rolling mills, and steel mills.

rail steel consumption grew and the rapidity with which that type of consumption exceeded the total steel production of a given year. This feature is brought forward in Table 3.4. Table 3.4 shows that

TABLE 3.4

TIME REQUIRED FOR NONRAIL CONSUMPTION OF DOMESTIC STEEL TO EXCEED TOTAL PRODUCTION OF STEEL

1 Year to Which Nonrail Consumption Applies	2 Nonrail Consumption of Steel (thousands of net tons)	3 Nearest Year in Which Total Steel Production Fell Short of Nonrail Consumption	4 Time Lag in Years (Col. 1 − Col. 3)
1871	38	1869	2
1872	52	1869	3
1873	76	1869	4
1874	76	1869	5
1875	105	1871	4
1876	126	1871	5
1877	144	1871	6
1878	180	1872	6
1879	256	1873	6
1880	290	1874	6
1881	230	1873	8
1882	275	1874	8
1883	393	1874	9
1884	458	1875	9
1885	683	1877	8
1886	848	1878	8
1887	1,027	1878	9
1888	1,466	1880	8
1889	1,855	1881	8
1890	2,394	1885	5

Column 2: Taken from Table 3.2, Column 3.
Column 3: Taken from Table 3.2; A.I.S.A., *Annual Report, 1889*, p. 63.

the time required for the nonrail consumption of steel to exceed the total production of a given year varied from two to nine years, the average being about six years. Consequently, if the nonrail demand for steel was inelastic over the range of prices involved, the observed scale of operations could have been achieved with an average lag of six years even in the absence of rails.

It is possible to estimate the maximum gain to the nation made possible by rail-induced economies of scale. The computation turns on the availability of the open-hearth furnace. The optimum plant size of the open-hearth mill was about one tenth that of a Bessemer mill.[30] In 1880, open-hearth mills had an average production of only 3,500 net tons; in 1890, the average product was 9,000 net tons.[31]

If the absence of rails reduced the scale of operation to such a level that the cost of Bessemer steel exceeded the cost of open-hearth steel, consumers of Bessemer steel would have made their purchases from open-hearth mills. Hence the maximum gain to society from the economies of scale in Bessemer plants induced by rails was the price differential between Bessemer and open-hearth steel. In 1880, the average differential in the delivered price of Bessemer and open-hearth steel was $10.19 per net ton. Nonrail consumption of Bessemer steel was 112,200 net tons. Hence the maximum gain due to economies of scale induced by the railroad in the production of nonrail steel was only $1,144,000, or 0.01 per cent of gross national product. By 1890, the average price differential fell to $4.02 per ton, while the production of nonrail Bessemer steel rose to 1,740,000 net tons. The indicated maximum gain attributable to railroad-induced economies of scale in 1890 is thus $6,995,000, or 0.06 per cent of gross national product.[32]

It appears that a modern steel industry would have emerged even in the absence of a demand for rails. While the increase in the scale of operations may have lagged behind that actually observed, it does not appear likely that the average lag would have exceeded six years. In any case, the maximum social loss of a slower growth in the scale of operations would have been barely one twentieth of 1 per cent of gross national product. Perhaps the most striking difference in the first quarter century of the development

---

[30] Peter Temin, *Iron and Steel in Nineteenth-Century America* (Cambridge: M.I.T. Press, 1964), p. 141.

[31] *Tenth Census, Manufacturing*, pp. 743, 756; Temin, *op. cit.*, p. 174; *Eleventh Census, Manufacturing*, Part III, p. 412.

[32] A.I.S.A., *Annual Report, 1896*, p. 69; *Eleventh Census, Manufacturing*, Part III, p. 417; *Historical Statistics*, p. 139. Even the small figures indicated in this computation overstate the gain attributable to railroad-induced economies of scale. The computations assume that the superior quality of open-hearth steel was of no value to consumers of the Bessemer product. One would expect that for at least some purchasers of Bessemer steel there was a positive value to the superior features of open-hearth steel, although they obviously believed that the incremental benefit was less than the price differential between the two products. The computations also assume the existence of a completely inelastic demand curve for nonrail Bessemer steel. However, if the demand curve had some degree of elasticity, then to some consumers the incremental value of steel over other alternatives was less than the observed differential in the price of Bessemer and open-hearth steel.

of a modern steel industry in the United States would have been the predominance of open-hearth over Bessemer mills.

The Effect of Railroads on Technological Innovations

In discussing the effect of railroads on the introduction and diffusion of inventions, a distinction has to be drawn between innovations limited exclusively, or largely, to the operation of railroads and innovations that had a major impact outside of the railroad industry. The former category of devices will be denoted by the term "restricted," the latter by the term "transcending."

Most inventions which arose out of the operation or construction of railroads fall into the category of restricted devices. Such items as air brakes, block signals, car trucks, automatic-coupling devices, track switches, pullman cars, and equalizing bars were mere appurtenances. While they may have been important to the efficient operation of railroads, they had no significant application outside of this industry during the nineteenth century. Nor did the railroads' demand for these items induce the rise of industries or production processes of transcending economic significance. They made no independent contribution to economic growth. Rather they defined the conditions under which railroads operated — conditions that in varying degrees explain how it was that railroads were able to produce low-cost transportation service.

There are, however, certain innovations associated with railroads to which transcending significance has been attached, and which therefore require further consideration. The two that will be considered here are cheap methods of producing steel, and the telegraph.

The belief that the railroads' demand for improved rails was responsible for the inventions that led to low-cost steel production rests on shaky foundations. It was not the problem of how to produce better rails that led Henry Bessemer into the series of experiments that eventuated in the Bessemer converter. As J. S. Jeans has pointed out, Bessemer's experiments stemmed from a desire to improve the effectiveness of artillery. A major obstacle to such improvement was the inadequacy of cast iron for cannon firing heavy projectiles. Bessemer pursued his research on metallurgy with the aim of "producing a quality of metal more suitable than any other for the construction of heavy ordnance." [33] William Kelly, who in-

[33] J. S. Jeans, *Steel: Its History, Manufacture, Properties and Uses* (London, 1880), pp. 44–45.

dependently discovered the Bessemer process in the United States, was engaged not in the production of heavy rolled forms but in the production of cast-iron kettles.[34] And the immediate factor that stimulated the Siemens brothers to develop the open-hearth furnace was their desire to find industrial applications for their earlier discovery, the regenerative condenser.[35]

The point is that metallurgical research in the mid-nineteenth century was induced by the rapidly growing demand for iron in a wide variety of production processes. Cheap steel had potential marketability not only for the fabrication of rails but also for ships, boilers, bridges, buildings, ordnance, armor, springs, wire, forgings, castings, chains, cutlery, etc. Metallurgical innovators could have been, and were, lured to search for improved products and processes by the profit that was to be earned from sales in all markets for iron, and not just in the railroad market.

It is, of course, true that during the initial decades following the discovery of the Bessemer process, most of the steel that poured from converters was destined for the fabrication of rails. From 1867 to 1883, about 80 per cent of all Bessemer steel ingots were so consumed.[36] This fact seems to suggest that railroads played an essential role in making cheap steel available on a large scale.

The problem is, however, more complex than it first appears. The fact that rails used four fifths of Bessemer steel production — in some years the share approached or exceeded 90 per cent — raises the question of whether the Bessemer process was, during this period, a restricted or a transcending innovation. If Bessemer steel had been used exclusively for railroad purposes one could give an unequivocal answer to the question. Like block signals or air brakes it would fall into the category of an appurtenance that contributed to economic growth only in and through the railroads; Bessemer steel would have increased real national income only because it increased productivity in the railroad industry.

But in fact some Bessemer steel was used for nonrail products. In 1880, 8.9 per cent of all such metal was turned into bars; 5.7 per cent into rods; 0.2 per cent into structural shapes, sheets, and boiler or other plate. In total, 169,645 net tons of Bessemer steel were rolled into products other than rails.[37] While some of these

[34] *The National Cyclopaedia of American Biography* (46 vols., New York, 1898–1963), Vol. XIII, p. 196.
[35] Leslie Stephen and Sidney Lee (eds.), *The Dictionary of National Biography* (22 vols.; Oxford, 1921–1922), Vol. XVIII, pp. 241–242.
[36] A.I.S.A. *Annual Report, 1889*, pp. 63, 65; Note to Table 3.2, Column 2.
[37] *Ibid.*; also *Tenth Census, Manufacturing*, pp. 743, 758.

shapes were also consumed by railroads, it may be assumed that the bulk was not. Was the amount of Bessemer steel used for non-railroad purposes large enough to warrant the classification of the Bessemer process as a transcending innovation?

While a definitive answer to this question requires more thorough research than was possible for this paper, available data suggest that the tentative answer should be: "No." In 1880, Bessemer steel accounted for only a minuscule share of the most important non-rail forms of rolled ferrous metal. Bessemer steel accounted for 9.79 per cent of rolled bar, 0.57 per cent of structural shapes, and 0.50 per cent of sheets and plates.[38] At the time when Bessemer steel was most rapidly displacing wrought iron as the basic raw material for rails, it was unable to dislodge wrought iron from its dominance as an input in the production of other products of rolling mills. According to Peter Temin, Bessemer steel "was subject to mysterious breakages and fractures that made people prefer iron" for most nonrail purposes.[39]

In later years the fall in the relative price of steel led to a situation in which most rolling-mill products were made from this material. By 1909, over 89 per cent of rolling-mill output other than rails was made from steel. However, most of this steel came not from converters but from open hearths. In the production of those forms for which demand was increasing most rapidly — plates, sheets, structural shapes, wire, etc. — open-hearth steel was preferred because of its superior qualities and its competitive price.[40]

Thus superior alternatives to Bessemer steel for nonrail uses were available at comparable prices through the end of the nineteenth century. Bessemer steel does not appear to have provided the basis for the remarkable growth of productivity in other industries that it did in the railroad sector.[41] Although the rapid expansion of Bessemer production between 1867 and 1890 is attributable to the

[38] *Tenth Census, Manufacturing*, pp. 754, 755, 758.
[39] Temin, *op. cit.*, p. 217.
[40] *Ibid.*, pp. 141–145, 217, 224–230, 279–280. It should be stressed that only a negligible proportion of open-hearth steel was used for the production of rails. In 1880 the figure was 10 per cent, by 1890 it had fallen to less than 1 per cent.
[41] According to Albert Fishlow ("Productivity and Technological Change," *op. cit.*) the substitution of steel rails for iron rails, the increased power of locomotives, and the increased capacity of freight cars account for 50 per cent of the increase in total factor productivity between 1870 and 1910. He further indicates that steel rails were a necessary condition for the utilization of heavy engines. They may also have been a necessary condition for the larger capacity of freight cars.

market for rails, the Bessemer process appears to have been a restricted rather than a transcending innovation.

Unlike the case of Bessemer steel there was little connection between the railroads and the early growth of the telegraph industry in the United States. The first telegraph line was established in 1844. Just eight years later the nation was laced with lines totaling 17,000 miles.[42] By 1852, the telegraph network connected all of the major eastern and southern cities. St. Louis, Milwaukee, Chicago, Detroit, and Toledo had telegraphic connection with the Atlantic coast well before the completion of the railroad link. In the United States, said the Superintendent of the Census, "the telegraphic system is carried to greater extent than in any other part of the world, and the numerous lines now in full operation form a network over the length and breadth of the land. They are not confined to the populous regions of the Atlantic coast, but extend far into the interior, climb the sides of the highest mountains, and cross the almost boundless prairies." [43]

The demand that induced the remarkably rapid rate of construction emanated not from railroads but from other businesses. Bankers, stock brokers, commodity brokers, and newspapers were the largest purchasers of telegraphic service.[44] The railroads did not try to employ the new device systematically until the basic wire network had been completed. The Erie Railroad, which began to use the telegraph for the dispatching of trains in 1851, was the first to do so. Despite its successful experience, most other railroads did not immediately become convinced of the advantages of the system. As late as 1854, a reporter for the *London Quarterly Review* could write that "the telegraph is rarely seen in America running beside the railway." [45]

What then is the basis for the view that railroads played a fundamental role in the growth and diffusion of the telegraph? It appears to rest primarily on the operating efficiencies achieved as a result of the alliance between the two systems. Once the alliance was formed, telegraph companies could utilize railroad men "to watch the line, straighten poles, re-set them when down, mend wires and report to the telegraph company." [46] Consequently, by making use

---

[42] U.S. Bureau of the Census, *Report of the Superintendent of the Census for December 1, 1852* (1853), pp. 111–113.

[43] *Ibid.*, p. 106.

[44] Robert Luther Thompson, *Wiring a Continent* (Princeton: Princeton University Press, 1947), pp. 47, 242.

[45] *Ibid.*, pp. 203–210.

[46] *Ibid.*, p. 213.

of railroad trackwalkers and other railroad maintenance men, telegraph companies were able to reduce maintenance costs below what they would have been. Such savings were, no doubt, reflected in lower rates that probably stimulated the growth of telegraphic service. Given the existence of railroads, the telegraph was no doubt aided by the alliance. However, it by no means follows that the observed rate of growth of the telegraph industry was higher than it would have been in the absence of railroads. It must be remembered that by speeding up the distribution of mail, railroads provided consumers with a better substitute for the telegraph than would otherwise have been available. Consequently, whether railroads made the volume of telegraphic business larger or smaller than it would have been in their absence is at this point a moot question.

The Social Saving of Railroads[47]

It is possible to set an upper limit on the increase in national income attributable to the reduction in transportation costs made possible by railroads. The main conceptual device used in this computation is the "social saving." The social saving in any given year is defined as the difference between the actual cost of shipping goods in that year and the alternative cost of shipping exactly the same goods between exactly the same points without railroads. This cost differential is in fact larger than the "true" social saving. Forcing the pattern of shipments in a nonrail situation to conform to the pattern that actually existed is equivalent to the imposition of a restraint on society's freedom to adjust to an alternative technological situation. If society had had to ship by water and wagon without the railroad, it could have altered the geographical locus of production in a manner that would have economized on transport services. Further, the sets of primary and secondary markets through which commodities were distributed were surely influenced by conditions peculiar to rail transportation; in the absence of railroads some different cities would have entered these sets, and the relative importance of those remaining would have changed. Adjustments of this sort would have reduced the loss of national income occasioned by the absence of the railroad.

For analytical convenience the computation of the social saving is divided into several parts. We begin with the estimation of the social saving on the interregional distribution of agricultural com-

[47] Unless otherwise stated, Fogel, *op. cit.*, Chapters 2 and 3, is the source for all of the computations presented in this section.

modities. In 1890, most agricultural goods destined for interregional shipment were first concentrated in the eleven great primary markets of the Midwest. These farm surpluses were then transshipped to some ninety secondary markets located in the East and South. After arriving in the secondary markets the commodities were distributed to retailers in the immediately surrounding territory, or exported.

Of the various forms of transportation in use in 1890, the most relevant as an alternative to railroads were the waterways. All of the eleven primary markets were on navigable waterways. Lakes, canals, rivers, and coastal waters directly linked the primary markets with secondary markets receiving 90 per cent of the interregional shipments. Consequently it is possible to compute a first approximation of the interregional social saving by finding the difference between payments actually made by shippers of agricultural products and the payments they would have made to water carriers if shippers had sent the same commodities between the same points without railroads.

The agricultural tonnage shipped interregionally, in 1890, was approximately equal to the local deficits of the trading regions of the East and South plus net exports. The local net deficits of a trading area are computed by subtracting from the consumption requirements of the area its production and its changes in inventories. The average rail and water distances of an interregional shipment are estimated from a randomly drawn sample of the routes (pairs of cities) that represent the population of connections (i.e. all possible pairings) between primary and secondary (deficit) markets. The water and rail rates per ton-mile for the various commodities are based on representative rates that prevailed in 1890 over distances and routes approximating the average condition. The application of observed water rates to a tonnage greatly in excess of that actually carried by waterways is justified by evidence which indicates that water transportation was a constant or declining-cost industry.

Using these estimates of tonnages shipped, rates, and distances it appears that the actual cost of the interregional agricultural transportation in 1890 was $87,500,000, while the cost of transporting the same goods by water would have been only $49,200,000. In other words, the first approximation of the interregional social saving is negative by about $38,000,000. This odd result is the consequence of the fact that direct payments to railroads included virtually all of the cost of interregional transportation, while direct payments to water carriers did not. In calculating the cost of shipping without the railroad one must account for six additional items of cost not included in payments to water carriers. These items are cargo losses in transit, transshipment costs, wagon-haulage costs from water points to secondary markets not on waterways, capital costs not reflected in water rates, the cost resulting from the time lost when using a slow medium of transportation, and the cost of being unable to use water routes for five months out of the year.

The first four of the neglected costs can be estimated directly from available commercial data. Insurance rates measure the average cargo loss per dollar of goods shipped by water. Transshipment rates were published. Data on the capital invested in the construction and improvement of waterways were also published. The quantity of goods required by secondary markets not on waterways is indicated in the calculation of the net deficits of trading areas. And estimates of the cost of wagon transportation are available.

It is more difficult to determine the cost of the time lost in shipping by a slow medium of transportation and the cost of being unable to use water routes for about five months during each year. Such costs were not recorded in profit-and-loss statements, the publications of trade associations, the decennial censuses, or any of the other normal sources of business information. Consequently, they must be determined indirectly through a method that links the desired information to data which are available. The solution to the problem lies in the nexus between time and inventories. If entrepreneurs could replace goods the instant they were sold, they would, *ceteris paribus,* carry zero inventories. Inventories are necessary to bridge the gap of time required to deliver a commodity from its supply source to a given point. If, on the average, interregional shipments of agricultural commodities required a month more by water than by rail, and if water routes were closed for five months out of each year, it would have been possible to compensate for the slowness of water transportation and the limited season of navigation by increasing inventories in secondary markets by an amount equal to one half of the annual receipts of these markets. Hence, the cost of the interruptions and time lost in water transportation is the 1890 cost of carrying such an inventory. The inventory cost comprises two elements: the foregone opportunity of investing the capital represented in the additional inventory (which is measured by the interest rate), and storage charges (which were published).

When account is taken of the neglected costs, the negative first approximation is transformed into a positive social saving of $73,000,000 (see Table 3.5). Since the actual 1890 cost of shipping the specified commodities was approximately $88,000,000, the absence of the railroad would have almost doubled the cost of shipping

TABLE 3.5

The Social Saving in the Interregional Distribution of Agricultural Commodities

First approximation	$ −38,000,000
Neglected cargo losses	6,000,000
Transshipping	16,000,000
Supplementary wagon haulage	23,000,000
Neglected capital costs	18,000,000
Additional inventory costs	48,000,000
Total	$ 73,000,000

agricultural commodities interregionally. It is therefore quite easy to see why the great bulk of agricultural commodities was actually sent to the East by rail, with water transportation used only over a few favorable routes.

While the interregional social saving is large compared to the actual transportation cost, it is quite small compared to annual output of the economy — just six tenths of 1 per cent of gross national product. Hence, the computed social saving indicates that the availability of railroads for the interregional distribution of agricultural products represented only a relatively small addition to the production potential of the economy.

The estimation of the social saving is more complex in intraregional trade (movements from farms to primary markets) than in long-haul trade. Interregional transportation represented a movement between a relatively small number of points — eleven great collection centers in the Midwest and ninety secondary markets in the East and South. But intraregional transportation required the connection of an enormous number of locations. Considering each farm as a shipping point, there were not 11 but 4,565,000 interior shipping locations in 1890; the number of primary markets receiving farm commodities was well over 100.[48] These points were not

[48] In the intraregional case the term "primary markets" refers not merely to the eleven great midwestern collection centers but also to cities that served as collection centers for intraregionally traded commodities. Thus, while New York City was a secondary market for the corn, wheat, beef, and pork of the North Central states, it was a primary market for the dairy products, fruits, and other commodities produced by local farmers.

all connected by the railroad network, let alone by navigable waterways. The movement of commodities from farms to primary markets was never accomplished exclusively by water or by rail. Rather, it involved a mixture of wagon and water or wagon and train services.

A first approximation of the intraregional social saving (α) can be computed on the basis of the relationship shown in Equation 1:

$$\alpha = x[w(D_{fb} - D_{fr}) + (BD_{bp} - RD_{rp})] \qquad (1)$$

where

- x = the tonnage of agricultural produce shipped out of counties by rail
- w = the average wagon rate per ton-mile
- B = the average water rate per ton-mile
- R = the average rail rate per ton-mile
- D_{fb} = the average distance from a farm to a water shipping point
- D_{fr} = the average distance from a farm to a rail shipping point
- D_{bp} = the average distance from a water shipping point to a primary market
- D_{rp} = the average distance from a rail shipping point to a primary market

The first term within the square bracket $w(D_{fb} - D_{fr})$ is the social saving per ton attributable to the reduction in wagon transportation; the second term $(BD_{bp} - RD_{rp})$ is the social saving per ton on payments to water and rail carriers. One of the surprising results is that only the first term is positive. In the absence of railroads, wagon transportation costs would have increased by $8.92 for each ton of agricultural produce that was shipped intraregionally by rail. However, payments to water carriers would have been $0.76 per ton less than the payments to railroads. In other words, the entire first approximation of the α estimate of the social saving — which amounts to $300,000,000 — is attributable not to the fact that railroad charges were less than boat charges but to the fact that railroads reduced the amount of expensive wagon haulage that had to be combined with one of the low-cost forms of transportation.

To the $300,000,000 obtained as the first approximation it is necessary to add certain indirect costs. In the long-haul case the first approximation of the social saving omitted six charges of considerable importance. In the intraregional case, however, three of these items were covered by the first approximation. Wagon-haulage costs are included in Equation 1. Transshipment costs would have been no greater in the nonrail case than in the rail case. In both situations bulk would have been broken when the wagons reached the rail or water shipping points, and no further transshipments would have been required between these points and the primary markets. Since all government expenditures on rivers and canals financed out of taxes rather than tolls were assigned to interregional agricultural shipments, their inclusion in the intraregional case would represent double counting.

Three indirect costs do have to be added to the first approximation. These are cargo losses, the cost of using a slow medium of transportation, and the cost of the limited season of navigation. As is shown by Table 3.6 these neglected items amount to only $37,000,000, which, when added to the first approximation, yields a preliminary α estimate of $337,000,000, or 2.8 per cent of gross national product.

TABLE 3.6

The Preliminary Intraregional Social Saving
(in millions of dollars)

First approximation	$ 300.2
Cargo losses	1.3
Cost of slow transportation	1.7
Cost of the limited season of navigation	34.0
Total	$ 337.2

The preliminary estimate of the agricultural social saving is based on the severe assumption that in the absence of railroads all other aspects of technology would have been unaltered. It seems quite likely, however, that in the absence of railroads much of the capital and ingenuity that went into their perfection and spread would have been turned toward the development of other cheap forms of land transportation. Under these circumstances it is possible that the internal-combustion engine would have been developed years sooner than it actually was, thus permitting a reduction in transportation costs through the use of motor trucks.

While most such possibilities of a speed-up in the introduction and spread of alternative forms of transportation have not been sufficiently explored to permit meaningful quantification at the present time, there are two changes about which one can make fairly definitive statements. These are the extension of the existing system of internal waterways, and the improvement of common roads. Neither of these developments required new knowledge. They merely involved an extension of existing technology.

As has already been pointed out, in the absence of railroads it would have been technologically and commercially feasible to build canals in the North and South Central states. A five-thousand-mile system of such canals would have brought all but 7 per cent of agricultural land within forty "airline" miles of a navigable waterway. In so doing, these waterways would have reduced the combined inter- and intraregional social saving from $410,000,000, to $287,000,000. Similarly, improvement of roads could have cut the cost of wagon transportation by 39 per cent, thus still further reducing the total agricultural social saving to $214,000,000.

It is possible to estimate roughly the social saving on nonagricultural commodities by extrapolating the social saving per ton-mile on agricultural commodities to nonagricultural commodities.[49] To do so, two adjustments must be made. First, the figure of $214,000,000 presented as the agricultural social saving already includes substantial elements of the social saving on nonagricultural items. Although all of the capital costs of the improvement of waterways were charged to agricultural commodities, most of this cost should be distributed among nonagricultural items. Similarly the wagon rates used in the computations assumed zero return hauls so that these rates cover most of the additional wagon cost that would have been incurred in shipping nonagricultural commodities to farms. It is probable that 35 per cent of the $214,000,000 should be assigned to the social saving induced by railroads in transporting products of mines, forest, and factories. In other words, the "pure" agricultural social saving is about $140,000,000. Given that 20,000,000,000 ton-miles of railroad service were required for the shipment of agricultural commodities in 1890, the agricultural social saving per ton-mile of railroad service was $0.0070.

Second, available evidence suggests that the social saving per ton-mile was less on nonagricultural products than on agricultural ones. Products of mines dominated the nonagricultural commodities carried by railroads. Coal alone represented 35 per cent of the non-

[49] This extrapolation involves a series of uncertain assumptions. The hazards involved in such an extrapolation are discussed in Fogel, *op. cit.*, pp. 219–224. A reliable estimate of the total social saving can only be obtained by applying to nonagricultural commodities the detailed methods of estimation used in the agricultural case. It seems probable that such a computation will reveal that the social saving on all freight was well below 5 per cent of gross national product (cf. *ibid.*, p. 223). However, pending such a computation, estimates of the social saving on all freight — including the one given in this chapter — should be considered tentative and largely illustrative.

agricultural tonnage in 1890. Iron and other ores brought the mineral share to over 50 per cent. As has previously been shown, a relatively small extension of the canal system could have brought most mines into direct contact with waterways. Thus, very little supplementary wagon transportation would have been required on these items. Moreover, the cost of increasing inventories to compensate for the slowness of, and interruptions in, water transportation would have been quite low. The total value of products of mines was well below the value of the agricultural commodities shipped from farms. As a consequence, the opportunity cost of the increased inventories of minerals would have been well below that found for agriculture. Additional storage charges, if any, would have been trivial. Minerals required neither very expensive storage facilities nor shelters. They were stored on open docks or fields.

These general considerations are supported by estimates compiled by Albert Fishlow, which reveal that in 1860 the social saving per ton-mile on nonagricultural commodities was only 46 per cent of that for agricultural goods.[50] This ratio applied to the figure of $0.0070 indicates that in the case of nonagricultural freight the social saving per ton-mile in 1890 was $0.0032. On this basis the social saving made possible by the 59,000,000,000 ton-miles of nonagricultural freight service provided by railroads in 1890 was $189,000,000. The last figure added to the "pure" agricultural saving yields a total of $329,000,000 on all commodities. Thus, the availability of railroads for the transportation of commodities appears to have increased the production potential of the economy by about 3 per cent of gross national product.

Implications for the Space Effort[51]

If space activities should make a major contribution to economic growth it seems highly unlikely that this will take the same form

[50] Albert Fishlow, *Railroads and the Transformation of the Ante Bellum Economy* (forthcoming, Harvard University Press, Fall, 1965), Chapter 2; Fishlow, "The Economic Contribution of American Railroads Before the Civil War," unpublished doctoral dissertation, Harvard University, 1963, Table A-5. Fishlow estimates that the social saving on passenger traffic in 1890 was $300,000,000 (see Fishlow, *Railroads*, Chapter 2).

[51] It is worth repeating the caveat stated at the beginning of this chapter. No evaluation of the impact of railroads on American development can be complete without a consideration of the cultural, political, military, and social consequences of such an innovation. I have focused on economic aspects of railroads, and my consideration of the implications of the railroad experience for the space effort is limited to the economic effects outlined in the preceding pages.

as the contribution made by railroads. The central feature of the developmental impact of the railroads was not so much that they induced or made possible new economic activities but that by reducing transportation costs they facilitated processes and activities which were well under way prior to their advent. The cheap form of inland transportation service purveyed by railroads speeded the commercialization of agriculture, widened the market for manufactured goods, and promoted regional specialization. While the scope of these effects and the net benefit accruing to the economy from them was much less than is usually presumed, they were nonetheless large enough to warrant the investment made in railroads.

The present cost of rocket transportation between points on earth is now far more expensive than alternative services. It is conceivable that future improvements may reduce the cost of rocket transportation to such a level that the movement of persons between the most distant points on earth will become commercially feasible.[52] But experience thus far offers no basis for assuming that rockets will, in the foreseeable future, provide a superior alternative to boats, trains, trucks, pipelines, and planes for the transportation of the freight now normally carried by these mediums. Moreover, since the development of supersonic jets will soon reduce the flying time between the most distant points on earth to about six hours, the maximum time saving in earth transportation that could be brought about by rockets is only about five hours.

As for access to resources, it does not seem likely that rockets will match the quite limited contribution of railroads to the expansion of the available supply of traditional economic resources. Even if other planets of the solar system have agricultural lands several times more productive than the best prairie soils, or ore deposits richer than the Mesabi mines in their heyday, the cost of transporting the products of these planets to earth would, under existing circumstances, be prohibitive. The mining of planetoids or other extraterrestrial bodies, as has sometimes been proposed, would therefore become practical only if there were a very sharp decrease in the cost of space transportation or a very sharp rise in the cost of available resources on earth. Considering the substitutability of materials and the rapid rate of innovation in synthetics, the likelihood of a contribution in this area during the foreseeable future seems very small.

[52] Leston Faneuf, "Application of Space Science to Earth Travel," *Peacetime Uses of Outer Space*, ed. Simon Ramo (New York: McGraw-Hill, 1961), pp. 86–109.

With respect to the development of technology, the experience of the railroads offers little basis for assuming that the devices required for space transportation will lead to innovations that markedly affect productive techniques in other spheres of activity. Most of the devices invented to improve railroads had no significant applications outside of this industry. And in the case of Bessemer steel, a product with many applications that was clearly promoted by railroad requirements, the economy independently produced an extremely effective substitute, namely, open-hearth steel.

It is also important to keep in mind that the huge investment devoted to railroad construction may, by diverting resources, have retarded technological development in other fields. A case in point is the relatively late realization of commercially useful motor vehicles. The decisions that led society to invest billions of dollars in railroads between 1830 and 1890 while allocating paltry sums to the perfection of motor vehicles may have delayed the advent of motor transportation by several decades.

Given the information generally available in 1830, the preference for railroads over the horseless carriage is quite understandable. Although carriages powered by steam engines were sufficiently perfected to be commercially successful in the absence of railroad competition, they were less efficient than the early railway.[53] The experience gained as the railroad system expanded revealed specific ways in which locomotives, cars, and track could be improved. With millions of dollars invested in new railroads annually, and with virtually each new extension of the system embodying some technological advance, the efficiency gap between the railroad and the horseless carriage became increasingly great — perhaps to the point that the latter no longer appeared to be a practical alternative. At any rate, it is interesting to note that the pace of experimentation on horseless carriages and wagons does not appear to have regained the peak attained in the 1820's and 1830's until half a century later. By then it cost the Minnesota farmer as much to haul a ton of wheat 15 miles by wagon as it did to ship the same quantity 375 miles by rail. And the railroad network was too dense to reduce significantly the cost of wagon haulage by further additions to the system. Thus motor vehicles were perfected and put into commercial production when the developmental potential of railroads had more or less run its course.

[53] D. C. Field, "Mechanical Road-Vehicles," *A History of Technology*, eds. Charles Singer and Others (New York and London: Oxford University Press, 1958), Vol. V, pp. 420–426.

Could the horseless carriage have been made commercially viable sooner than it actually was? The crucial step in the perfection of motor vehicles appears to have been the substitution of the internal-combustion engine for the steam engine as the power source. However, "not only the fundamental internal-combustion engine theory, but even that of the diesel engine" was published as early as 1824.[54] Consequently, one cannot at this point foreclose the possibility that in the absence of railroads more capital and talent would have been devoted to the perfection of the horseless carriage, and that as a result the engineering knowledge and technical skills required to produce effective motor vehicles would have emerged decades sooner than they actually did. In weighing the benefits of technological innovations such as telemetry, compact power sources, reinforced plastics, and new metal alloys that have been induced or accelerated by the space effort, one must take into account alternative devices that could have been developed if the resources devoted to the space effort had been applied to other activities.

The preceding discussion does not necessarily imply that the space effort will fail to contribute to the expansion of the nation's production potential. There is a very important asymmetry between the railroads and rockets. While railroads increased the area of economically accessible territory, it did not affect physical accessibility. All parts of the earth's surface were penetrable before the invention of railroads. Indeed the various territories of the world were explored by man before they were traversed by rails. It is impossible, however, to explore the solar system without rocket ships. The development of spacecraft, unlike the railroad, thus offers man access to knowledge that cannot be obtained in any other way. The knowledge gained from space exploration may have enormous consequences for the biological and physical sciences. The advances in these sciences made possible by space exploration may in turn lead to a technological and commercial revolution far more portentous than that which followed from the scientific breakthroughs of the seventeenth, eighteenth, and nineteenth centuries. In this respect the space effort has a chance of affecting economic life far more radically than did the railroads. Examination of the railroad experience cannot aid in predicting the value of the scientific knowledge that may be gained from space exploration.

[54] Orville Charles Cromer, "Internal-Combustion Engines," *Encyclopaedia Britannica* (Chicago: 1961), Vol. 12, 494; Field, "Mechanical Road-Vehicles, *op. cit.*," pp. 426–434.

4

The Economic Impact of the Railroad Innovation

Paul H. Cootner

The railroad innovation was an important factor in shaping American economic development, but it was not nearly so important as some romantic historians have sometimes suggested. The difference between the role that it did play and the role that it did not is a difference that may seem overly subtle to some, but it is of crucial significance to students of technological change and economic development. In particular, these differences are especially important in estimating the pace and magnitude of economic change that might result from new space technology.

It is undeniable that large fractions of the output of important industries went to build the railroads and to operate them. It is also true that the rails carried major volumes of many essential raw materials at costs low enough to make their use profitable to producer and consumer alike. Large amounts of U.S. saving financed their construction, and thousands of workers were employed by them. Nevertheless, it is one thing to say that railroads were an essential part of the fabric of American economic development and quite another thing to say that they were the engine that drove it. What in fact happened was that the railroad became another of the economy's tools, one that was called upon only when the situation called for the tool in question. Thus, the rate of adoption of railroad technology varied in magnitude and in geography with the economy's pattern of demand for final goods, instead of being the exogenous force which stimulated the pace of economic development. It is on this point of difference that the argument of this paper rests, in the hope that it will enable us to perceive more clearly the relationship between our economy and present and prospective technological change.[1]

[1] Most of the argument of this paper is supported in more detail in Paul H. Cootner, "Transport Innovation and Economic Development: The Case of the

The Interaction of Railroads and the Economy

The Importance of Technological Competition

The important thing to remember is that the railroads were introduced into a sturdy, thriving economy in which aggressive and inventive entrepreneurs were constantly seeking out opportunities for personal gain in a framework that contemporary ideology believed would lead to the good of all. It was an economy that included not only the United States but also the United Kingdom, Europe, Asia, and South America, all tied together more or less tenuously by the bonds of trade. Tariffs and costs of transport impeded the flow of goods across national boundaries; but a shift in demand for cotton, sugar, or iron in one country shortly produced its effect in all of this trading area. Merchants sought to transfer goods whenever it was economic, and tried to uncover any means of lowering the costs of transfer.

One of these costs of transfer was land transportation, and one of the techniques for lowering these costs was the railroad. In this kind of economy the railroad was introduced wherever it was profitable as soon as it was profitable. Despite the aggrieved complaints of men who sought an even more rapid pace of development, there is little evidence of sluggishness or conservatism in that pace. The growth of the railroad net was retarded — not by apathy but by the *economic need* for railroad services. The pace of railroad construction was the product of several confluent forces. The first of these forces was the slow, continual, creeping improvement in rail transport technique that improved its competitive position vis-à-vis the canal, the river, the turnpike, the barge, and the coastal steamer. These victories were slow and hard fought, because each of the alternate ways had advantages over railways that made many transport decisions marginal ones. This feature of transport technology alone prevented railroads from exercising the dramatic causal role which is frequently assigned them.

This important point deserves elaboration. Rarely does a new technology make obsolescent all existing means of achieving a given end. To be sure, the innovation might permit some feat that was completely impossible previously, just as railroads made land travel at high speeds possible, and space innovations will presumably

---

U.S. Steam Railroad, 1826–1886," Ph.D. dissertation, Massachusetts Institute of Technology, 1953; and in "The Role of the Railroads in United States Economic Growth," *Journal of Economic History* (December 1963).

permit flights to the moon. But when we think in terms of the economic objective to be achieved, such as rapid communication, or tourism, or carriage of freight, it is a much rarer innovation that makes possible some desirable end that could not previously be accomplished at all. What the innovation usually involves is doing something better in some way (say faster) but worse in some other way (say more costly). The rate of adoption of the innovation in this case depends upon the need for doing things "faster" and the willingness to spend more money. Short of cataclysmic changes in final demand, however, the process is a gradual one.

Take the case of the railroad. It provided faster transportation for goods than did either canals or turnpikes. It was, however, especially at first, substantially more expensive to operate with bulk freight than water transport over level ground. On the other hand, where hilly terrain was concerned it proved more economic than canals because of the need for locks. However, roads could be built over hilly terrain for lower capital costs than railroads, though with higher operating costs. Every early decision to build a railroad depended on careful calculations of relative costs, and the first railroads were built where they had some special advantage, such as leading to coal mines in the hills, or between cities with considerable passenger traffic interested in speed and comfort. In the decade ending in 1840, more canal than railroad mileage was completed.

As time went on, of course, smaller technological changes and economic developments tended to improve the position of the railroads relative to that of competing forms of transport, and after a while railroads gradually prevailed over canals and turnpikes. Even then, however, this advantage developed because of the steady reduction in the relative costs of iron and machinery that characterized the entire economy rather than because of major technological advances in the railroads. The crucial test of imagination is to think about what would have happened to the U.S. economy in the absence of the railroad. One careful study by Robert Fogel [2] suggests that with extensive canal building the rate of development would not have been too different, with, of course, considerable changes in the pattern of trade and reduced economic activity in hilly and mountainous areas and greater activity along water courses. Whatever one feels about Fogel's specific finding, the major point is worth making: to the extent that railroads reduced the costs of economic development they speeded and amplified it, but the prime engine

[2] Robert William Fogel, *Railroads and American Economic Growth: Essays in Econometric History* (Baltimore: The Johns Hopkins Press, 1964).

that drove the U.S. economy was the goods it would produce and the ingenuity it would bring to bear in producing them.

The Pattern of Final Demand and Railroad Construction

In addition to the constraint on the assimilation of an innovation posed by competing methods of supplying the same service, there was another constraint posed by the shifting demand among industries that used the service. Unlike the popular picture, the railroad network did not spread smoothly from the settled area of the world into the unsettled. Instead, world railway construction tended to take place over long cycles, which alternated between the industrialized and the primary-producing regions, and the pattern of construction in the United States was merely the reflection of the worldwide pattern of demand in our national economy.

For many reasons, too complicated to detail here, nineteenth-century economic development took place in an irregular manner.[3] Instead of all facets of the economy proceeding at a smooth and steady pace, industrial growth proceeded at a rapid pace until the expansion pressed against the supply of raw materials. Rising prices for these raw materials set off a chain of investment in unexplored or undeveloped areas of the world. There was usually a long gap of time between the onset and the completion of these investments, and, until the fruit of these outlays began to appear, investment took place at a hectic pace. When the boom was over, supplies of raw materials had become available not only for the present but, at relatively little additional cost, for a number of years in the future as well. With cheap supplies assured, it was then possible for industrial growth to prosper exceedingly until raw material supplies became inelastic once more.

In the nineteenth century, the decades following 1810, 1830, and 1850 can each be identified as periods of high investment in primary goods production throughout the world. The alternate decades emphasized industrial investment. Even within a country like the United States, New England and Middle Atlantic industry tended to prosper in alternation with the agricultural West.

The very earliest railroads were built in the late 1820's just prior to a big agricultural and mining boom; and the construction of railway mileage in the predominantly agricultural United States progressed very rapidly, exceeding that of the United Kingdom, while

---

[3] Paul H. Cootner, "Social Overhead Capital and Economic Growth," in *The Economics of Take-Off into Sustained Growth*, ed. W. W. Rostow (London: Macmillan and Company, 1963).

at the same time an even greater mileage was being absorbed in canals. The bulk of this early construction went to aid the transport of agricultural goods, although some of it went to facilitate commerce between eastern cities, which benefited indirectly from the agricultural boom. In the following decade, however, the more industrialized areas built most of the railroads. *This* was the period of railway mania in Great Britain, some fifteen years after the railway was invented in that country. Similarly, in the United States, which is commonly considered to have been in a deep depression during that decade, considerable mileage was added in New England with its textile mills, and in Pennsylvania and New Jersey with their iron industry. Relatively little mileage was built in the trans-Appalachian region. The 1850's, like the 1830's, were once again a period of railway construction in less-developed areas. The alternation continued throughout the century.

The manner in which the railway network generally followed the pattern of investment also suggests that railway innovation did not have a very pronounced causal effect on the direction of investment in the century of its invention. The alternation started before the railroad invention and continued after it without interruption.

In short, the railroad was a tool, and like all human innovation it made economic life a little easier in a number of ways. It was more important than some inventions and less important than others. But the important business of nineteenth-century economic life was growing crops, building factories, and performing services. Railroads made these processes simpler in some places, so that economic activity became easier in those areas, and, by competition, harder in other areas. It is easy to see the successes, but the blighted growth, the prosperity that was slightly diminished or never achieved is harder to detect. The net effect of the railroad was positive, but it was never so positive that it made the difference between development of this country and stagnation.

I think that the lesson for space innovation is similar. To have a major effect on the economy it must do things better than the existing ways of doing them. Apparently, as a layman reads the news, it performs certain communications tasks better, and possibly cheaper, than existing methods. All things considered, however, it seems unlikely ever to become a major artery of cargo transportation, especially considering the relatively weak position of the airplane in this business sixty years after its invention. Nor does tourism seem likely to be a major source of demand, nor, in view of the difficulties, does colonization. Some investigations of earth itself — such as weather forecasting, geological research, or navigation — may be easier from the vantage point of space, and if done on some collaborative basis may even be cheap. But the important lesson to be learned is that the key to the pace and magnitude of the impact of innovation in this field is a study of the alternative costs of other means of accomplishing similar ends.

The Timing and Magnitude of the Aggregative Impact on the U.S. Economy

The first substantial approaches to the railroad innovation came with the utilization of a high-pressure steam engine in Trevithick's first locomotive in 1801. The date is significant, since Watt's patents expired in 1800, and (1) he chose not to work on locomotives and (2) opposed the use of high-pressure engines. However, interest in steam locomotion was strong enough so that once the obstacle of the patent was removed, experimentation began. Trevithick continued to work on the locomotive, and an 1803 version actually did some commercial haulage.

The high-pressure engine was, however, only one of the three innovations essential to an economical locomotive. The second invention, the multitubular boiler, was developed in 1815. The third, the direct connection between the piston and the driving wheel, came in 1820. During all of this period, inventors were developing technically satisfactory locomotives, but these engines did not prove as adequate in terms of economy. Despite the solution of the major technical problems by 1820, the first commercial locomotive was not built by Stephenson until 1825; the first really successful locomotive along modern lines was built in 1829.

The slowness of railroad development was clearly economic. At that time, at least, where canals could be constructed — on level ground with nonporous soil — such transport was cheaper. The horsepower required for carrying goods on water was much less than by rail, and the capital costs of railroads were about the same. Even after sharp declines in the cost of iron, copper, fuel, and capital, the main railway construction of the 1820's was concentrated on carrying coal over hillsides that made canals impossible or unusually expensive.

The first real "railway boom" in the United States came in the 1830's and was concentrated during the first major trans-Appalachian land boom. In the 1840's, activity was most heavily concentrated in England, France, New England, and the Middle Atlantic states, and in the 1850's huge railway building activity coincided with another surge of U.S. westward expansion. By the onset of the Civil War, the locomotive was sixty years old by technical standards and at least thirty-five years old by commercial standards. Yet in the decade of the 1850's, railroad investment accounted for only 2.2 per cent of gross national product and only 13 per cent of a suspiciously low estimate of gross capital formation. If we take a very simple-minded estimate of iron-rail consumption as a fraction of total iron production we find that the railroad industry accounted for only 15 per cent of iron output in the 1850's, and Robert Fogel[4] has pointed out that this method of calculation overestimates the true effect of the industry.

For comparison's sake, nonfarm residential housing investment accounted for 4.2 per cent of GNP in the 1950's and 28 per cent of gross capital formation. The privately owned public utilities (electricity, gas, and telephone) accounted for 2 per cent of GNP and 12 per cent of gross private capital formation. Later, the automobile industry's share of steel output was to be about 20 per cent. To get a firmer grip on these numbers we should point out that the construction industry accounted for about as much steel as did automobiles — a demand that has little to do with innovation.[5]

TABLE 4.1

THE RAILROADS' SHARE OF GNP INVESTMENT

Periods	1 GNP (current prices)	2 Railroad Investment (current prices)	3 Railroad Investment (percentage of GNP)
1842–1851	2.12	.023	1.1
1852–1861	3.92	.085	2.2
1872–1881	8.40	.208	2.5
1882–1891	11.80	.246	2.1
1892–1901	15.00	.185	1.2

Column 1: 1842–1851, average of Gallman's figures for 1844 and 1849; 1852–1861, average of Gallman's figures for 1854 and 1859. Robert Gallman, "Gross National Product in the U.S., 1834–1909," paper delivered at National Bureau for Economic Research conference on income and wealth, September 5, 1963.
Column 2: 1842–1861, calculated from cost of construction estimates for 1842, 1850, and 1860. The 1842 estimate is derived by multiplying Poor's mileage in operation by $25,000. Henry V. Poor, *History of the Railroads and Canals of the United States of America* (N.Y.; J. H. Schultz and Co., 1860), Vol. I, p. 612. The 1850 and 1860 figures are from the *Tenth Census of the United States, 1880* (Washington, D. C., 1883).
Column 3: Column 2 divided by Column 1.

[4] Fogel, *op. cit.*
[5] Data are from survey published annually in the U.S. Department of Commerce, *Survey of Current Business.*

TABLE 4.2

RAILROADS' SHARE OF GNP INVESTMENT AND OUTPUT
(dollar items in billions)

Periods	1 Railroad Output (1929 prices)	2 Col. 1 as Percentage of GNP	3 Investment as Percentage of GNP	4 Total Percentage	5 Railroad Revenues Current Dollars	6 Col. 5 as Percentage of GNP	7 Investment as Percentage of GNP	8 Total Percentage
1832–1841					.004			
1842–1851					.020	0.9	1.1	2.0
1852–1861					.100	2.6	2.2	4.8
1862–1871					.240			
1872–1881					.500	6.0	2.5	8.5
1882–1891	1.061	4.7	2.1	6.8			2.1	
1892–1901	1.852	5.8	1.2	7.0			1.2	

Column 1: Taken from Melville Ulmer, *Capital in Transportation Communication and Public Utilities* (Princeton: Princeton University Press, 1960), see n. 49.
Column 2: Column 1 divided by the first column of Table 4.1.
Column 3: The fourth column of Table 4.1.
Column 4: The sum of Column 2 and Column 3.
Column 5: For 1832–1841 the average mileage in operation multiplied by a revenue estimate of $3,000 per mile. The figure for 1842–1851 is taken from Paul H. Cootner, "Transport Innovation and Economic Development: The Case of the U.S. Steam Railroad, 1826–1886," Ph.D. dissertation, M.I.T., 1953 (see n. 5). Revenues for 1849 were estimated at $30,500,000, from H. V. Poor, *History of the Railroads and Canals of the United States of America* (see n. 24); Ulrich B. Phillips, *History of Transportation in the Eastern Cotton Belt* (N.Y.: Macmillan, 1913), p. 405, see n. 31; and from the *American Railroad Journal*, XXII (1849). The figures for 1849 and 1850 were averaged. For 1852–1861, the U.S. Secretary of the Treasury estimated revenues at $91,000,000 for 1855, based on reports on individual railways adjusted for some reporting deficiencies. See John L. Ringwalt, *The Development of Transportation Systems in the United States* (Philadelphia, 1888), cited in n. 12. For 1862–1871, a gross-receipts tax of 2.5 per cent on all transportation in 1865, 1866, and 1867 implied railway revenues of $236,000,000; estimate is based on an average of these figures. For 1872–1881, an average of annual revenue figures is given in H. V. and H. W. Poor, *Manual of the Railroads of the United States, 1868/69–1886* (New York: H. V. and H. W. Poor, 1869–1886).

Tables 4.1, 4.2, and 4.3 show the timing and magnitude of the impact of the rail innovation. They all tend to show that the maximum direct impact of the innovation did not come until after the Civil War — at some point about fifty or more years after the first railroad was built. (A moment's thought will, moreover, show that this is not at all unusual for major innovations. The first atomic bomb was exploded in 1945, and the first atomic reactor produced commercial electricity in Hartwell, England, in 1955; but despite the great claims for atomic power in the postwar years it is unlikely

TABLE 4.3

RAILROAD INVESTMENT AND GROSS CAPITAL FORMATION

Periods	1 Gross Capital Formation (current prices)	2 Railroad Investment (percentage of GCF)	3 Railroad Investment (percentage of GCF II)
1842–1851	0.210	11	8
1852–1861	0.517	16	13
1862–1871			
1872–1881	1.73	12	
1882–1891	2.45	10	
1892–1901	3.45	5	

Column 1: Figures for 1842–1861 are taken from Robert Gallman, "Gross National Product in the U.S., 1834–1909," *op. cit.;* for 1872–1901, from U.S. Census, *Historical Statistics of the United States from Colonial Times to 1957* (Washington, D. C., 1960).
Column 2: The second column of Table 4.1 divided by the first column of this table.
Column 3: The third column of Table 4.1 (prior to rounding) divided by the third column of Table 3 in Gallman. That table in Gallman gives the ratio of gross national product (including home manufactures) in 1860 prices which Gallman calls GCF II. I have assumed that the same ratio holds in current prices.

that it will reach its maximum impact for at least two more decades.)

While this outline of the broad demand for economic resources is useful in tracing the development of railway demand, it is not necessarily the most important economic datum for our purposes. It would be interesting to measure the railroad impact not only by its use of resources but also by its contribution to the volume of output. Under normal economic conditions we expect such contributions to be measured by the value placed on railway services. (As we can see from Table 4.2 such railway revenues follow the same pattern in reaction to GNP that we ascribed to rail investment.) But several arguments have been raised in the economic literature which claim that these value measurements understate the true contribution of the railroad. These are all versions of the social overhead capital-external economies argument, and are made most cogently by Rosenstein-Rodan and Robert Fogel.[6] Basically, these

---
[6] Paul Rosenstein-Rodan, "Problems of Industrialization in Eastern and South Eastern Europe," *Economic Journal* (June/September 1943); and "Notes on the Theory of the Big Push," typescript, Center for International Studies, Massachusetts Institute of Technology, 1957; Robert W. Fogel, *The Union Pacific Railroad: A Case in Premature Enterprise* (Baltimore: The Johns Hopkins Press, 1960).

authors argue that the railroad made possible or profitable other economic activities, the benefits from which did not accrue to the railroads themselves. These are important questions on which the facts are not all in, but my most measured opinion is that, properly considered, there were no substantial discrepancies between the social and private products of railway investment.[7] It is, of course, true that a railway in Nebraska stimulated Nebraskan development, but it also impeded growth in competitive agricultural areas. If we look at economic growth from other than a nationalistic point of view, the value figures in the contribution to economic growth. But even if these figures do understate the magnitude of the impact, there is even less reason to believe that they misstate the timing of the impact: to believe, that is, that these external economies were greater in the pre-Civil War period than they were in the postwar period of westward expansion.

If the development of space technology follows the normal pattern of major innovations, it will be some time before it reaches its peak impact on the U.S. economy. It would be unwise to waste resources by making elaborate plans too long in advance of need for them and when the shape of future developments is still only dimly seen. There is a tendency, already visible in the civilian nuclear power program, to try to force the pace of development by extensive subsidy in premature preparation for a prospective need. Such prematurity raises the possibility that when effort is really needed the good will of the planners will have been exhausted.

The Effect of Railroad Expansion on Secondary Industries

Iron and Steel

Table 4.4 spells out one crude set of estimates of the effect of the steam railroad on the iron and steel industry. Another, more conservative estimate made by Fogel yields a smaller measure of the total impact. The important thing to observe in both of these estimates, however, is that increased railroad demand did not always mean increased use of domestic iron. Except for a brief period in the mid-1840's, the bulk of U.S. rail consumption before 1855 was supplied by imports, largely from Britain. It is by no means necessary that the secondary impact of an innovation should fall upon domestic industry. Except for the mid-1840's, when British iron production was largely diverted to internal expansion, it was cheaper

[7] Cootner, "The Role of the Railroads in United States Economic Growth," *op. cit.*; and "Social Overhead Capital and Economic Growth," *op. cit.*

TABLE 4.4
RAILROADS' USE OF IRON AND STEEL

Periods	1 Iron Rails Produced (000 tons)	2 Pig Iron Produced (000 tons)	3 Iron Rails Pig Iron (percentage)	4 Steel Rails Produced (000 tons)	5 Steel Rails Pig Iron (percentage)	6 Total Col. 3 + Col. 5 (percentage)	7 All Rail Imports (000 tons)
1828–1844	—*	(200)	0	0	0	0	(22)
1847	100	896	11	0	0	11	14
1849	24	728	3	0	0	3	69
1850–1855	71	642	11	0	0	11	229
1856–1860	181	830	22	0	0	22	136
1861–1865	274	907	30	0	0	30	67
1866–1870	512	1640	31	11	1	32	281
1871–1875	678	2518	27	139	6	33	297
1876–1880	409	2870	14	613	21	35	134
1881–1885	164	4821	3	1284	27	30	132
1886–1890	86	7929	1	1921	25	26	56

Column 1: For 1849–1890 figures taken from James M. Swank, *History of the Manufacture of Iron in All Ages* (2nd ed., Philadelphia, 1892), see n. 57; for 1847, from B. F. French, *History of the Rise and Progress of the Iron Trade in the U.S., 1621–1837* (New York: Wiley, 1858).
Column 2: Taken from U.S. Census, *Historical Statistics, op. cit.* The figure for 1828–1844 is annual average of production in 1828–1832, 1840, and 1842.
Column 3: Column 1 divided by Column 2, multiplied by 100.
Column 4: Taken from Swank, *op. cit.*
Column 5: Column 4 divided by Column 3, multiplied by 100.
Column 6: The sum of Column 3 and Column 5.
Column 7: Figures for 1856–1890 taken from Richard Rothwell, ed., *Mineral Industry* (New York: McGraw, 1892), Vol. I. Data for 1856–1863 refer to fiscal years. For 1840–1855, taken from French, *op. cit.*, "Railroad Iron Imports"; for 1828–1844, annual average imports calculated from a table of remitted duty (French, *ibid.*) by dividing by the duty rate of $22.40 per ton. See Frank Taussig, "The Tariff, 1830–1860, *Quarterly Journal of Economics*, II (1888), 314–346, 379–384.
* Negligible.

for the United States to purchase its requirements abroad. When the situation changed, after 1855, it was not due to aggregate demand at all, but to the shift in population and economic activity to the trans-Appalachian area, close to good coal deposits and to the newly discovered Lake Superior iron ore, which was protected by the best "tariff" of all, the considerable expense of land shipment across the mountains from the Atlantic coast. The critical factor was that the ore and coal were linked by *water* transport which was so cheap as to make it uneconomic to haul imported iron *by rail* into the Midwest.

The importance of this point is not to belittle the impact of the railroad on the iron and steel industry but rather to stress that its impact on the U.S. iron and steel industry did not automatically follow from the magnitude of its demands. Had the Lake Superior mines not been discovered, the railroad net would have been nearly as large and yet the portion of rail consumption supplied by U.S. industry would have been much smaller. The proper inference to be drawn is that projection of prospective impacts of innovation must be developed with careful regard for the relative economies of production.

This point can be driven home if we next look at the impact of the railway on locomotive production. The railroads played an even bigger role in the steam-engine industry. In 1838, there were 350 locomotives in operation in the United States, and 271 of them had been produced domestically. Together, they accounted for 7 per cent of total stock of steam horsepower. By 1849, locomotives accounted for 35 per cent of steam horsepower and by 1859 accounted for 60 per cent. Since railroad engines were smaller in size than most engines, they constituted an even bigger share of the number of engines. The striking part of the railroad's impact on the making of engines was not only the importance of the demand, but the extent to which — unlike the iron and steel industry — it met that demand by domestic production that soon led to a thriving export business. The unimportance of British competition after 1840 is largely the result of differences in capital endowment. Because capital was cheaper and traffic heavier in England, English locomotives were heavy and expensive, and suited for operation on excellent roadbeds. In the United States, traffic was light and could not justify carefully graded straight roads or large capital outlay for equipment. American railway builders showed more ingenuity in locomotive mechanical construction and more daring in the use of high-pressure steam for power plants. Since U.S. roads were more like those of other countries that were building railroads, many engines were exported. Here the situation was the reverse of that for iron and steel. The particular exigencies of the situation led to the United States becoming an exporter of locomotives and an importer of rails.

The Demand for Fuel

Despite the importance of coal as a railroad fuel in later years, the railroad had little effect on demand in that industry at the beginning. Anthracite was the most abundant form of coal available at the time railroad construction began, and while it was technically a satisfactory fuel its fire was so intense as to have seriously deteriorating effects on the materials used in the locomotive firebox, and thus to make it economically unsatisfactory. It was only as the cost of wood fuel continued to increase relative to that of coal, as more railroads were built in the bituminous coal areas beyond the Appalachians, and as the relative costs of firebox materials continued to fall, that coal was used in great amounts. Relatively little coal was used prior to the early 1850's, and it was only after the advent of cheap steel in the post-Civil War period that coal use became dominant. By that period, however, the rapid increase in total railroad fuel consumption and the growing share of that fuel demand supplied by coal made the railroads a major consumer of coal. By 1880, the railroads consumed some 15 to 20 per cent of total U.S. production, and they must have had a tremendous impact on the *growth* of coal consumption in the decade of the 1870's.

Railroads and the Organization of Economic Activity

Urbanization

There is, however, one way in which railroads stimulated the development of coal even before the 1870's. That was partly because the use of wood by the railroads tended to speed the exhaustion of fuel wood, which was in relatively limited and inelastic supply. But in a broader sense, the railroad stimulated the demand for coal fuel by its centralizing effect on economic production, and through it, upon urbanization. As population tended to an increasing extent to center itself in cities, its fuel needs had to be produced and transported from more distant points. In general, it proved cheaper to mine substantial quantities of coal from concentrated areas than to cut large quantities of wood from extensive forests. Also, as economic activity shifted from the hilly East to the level country west of the Appalachians, the relative availability of water power diminished and the demand for fossil fuel necessarily increased.

In the same fashion, though to a lesser degree, urbanization tended to increase the demand for iron and minerals as building products relative to wood, and to increase the trade required to import food into the urban areas and to export manufactures and foreign imports into the countryside. In this whole process it is hard to separate cause and effect. The railroad net made urbanization easier, and urbanization increased the demand for railroad services while shifting demand for individual products toward those items most suited for urban living.

"Mass versus Quality" Production

One of the less obvious, indirect impacts of the railroads was on the quality of production. The best example of this is documented by Louis Hunter with regard to the iron industry.[8] Prior to the improvement of transportation in trans-Appalachia, the prime source of manufactured metal products was the local blacksmith. The capital invested in a blacksmith shop was minimal, comprising rudimentary equipment and large inputs of human effort. In this region one of the main requisites for iron was that it should be easily workable with elementary tools. The malleability of such iron depended vitally upon its purity; thus, to ensure ease of local metal manufacture, iron was typically produced at a purity in excess of that required by the final product. This increased purity raised the cost of the raw material but lowered the "manufacturing" cost sufficiently to offset it.

As long as local demand was small and erratic and transport costs were high, the market was restricted enough to make it uneconomic to produce metal products in centralized factories. When transport costs were reduced, such factories *were* set up, and with more specialized machinery they not only reduced the costs of manufacture itself but permitted the use of iron of a quality more consistent with minimum structural requirements.

Two points are important to remember about this issue: first, that "lower quality" is not an invidious term; and second, that while railroad construction brought with it the lower transport costs that made mass production possible, the process would have taken place to some degree with increased population though without lower transport costs or with the same level of economic activity but with improvements in other transport techniques, such as canals or roads. Lower quality in this context does not mean shoddy, but merely represents a tradeoff between unnecessary care and lower cost. Nevertheless, the very important tendency toward large-scale production in the United States depended not only on a large land area under common nationality, as frequently asserted, but also on sufficiently low transport costs to make that area an *economic* unit.

The iron industry was not the only one in which large units became more common. Agricultural processing also became more concentrated, particularly in the nodal areas where rail and water

---

[8] Louis Hunter, "The Influence of the Market Upon Techniques in the Iron Industry of Western Pennsylvania up to 1860," *Journal of Economic and Business History*, I (1928), 241–281.

transportation networks met. Mining for iron ore, copper, and coal was done in larger units, and other metal processing also became more concentrated. Transport-cost reduction only increased the scale of production in those industries where the costs of large factories were substantially less than of small ones, but that did comprise an important segment of industry.

The Development of New Institutions

In another chapter in this book Alfred Chandler and Stephen Salsbury study the impact of the railroad on the technology of dealing with administrative and organizational problems, showing how the innovation shaped our ways of dealing with the problems of operating these large giants. These changes were quite important, but other institutional arrangements outside the rail industry itself were also modified by its growth.

The most important of these was that complex set of arrangements known as the "capital market." Briefly, a capital market is a "place" where funds, which in a private economy represent the control of physical resources, can be channeled into a productive enterprise. The term "place" is in quotes advisedly, since the capital market is not confined to any physical locality. It is a "set of arrangements" whereby large needs for funds can be satisfied by relatively small individual investments by people who have no entrepreneurial connection with, or operating knowledge of, the project being financed. Such a market is essential to the smooth flow of operations of large enterprises without extensive government intervention.

In the beginning, such markets were fragmentary in the United States, and many enterprises drew on European, particularly British, capital markets. But foreign capital was usually much more conservative than domestic capital, generally because of the obstacles to knowledge about domestic opportunities, and when it financed new railroad enterprises in the United States it was usually through the mediation of some state borrowing. There were the beginnings of such markets already — particularly in New England — but from the beginning of the railroad era until quite late in the nineteenth century, railroad securities and railroad financing dominated the market. A textile plant or a blast furnace might be financed by soliciting friends or selected wealthy men, but railroads were enterprises too big for such *ad hoc* techniques. As late as 1900 the only good source of data on long-term interest rates was supplied by quotations on high-grade railroad bonds.

The important thing about an institution like a capital market is that, like several of those to be discussed later, it rarely atrophies once it has developed, but rather has its usefulness channeled into other directions. As the capital needs of the railroads subsided, the market was called upon to finance other industries and government deficits, enabling developments that were not foreseen when the market began to ripen more easily.

Another such institution was technical education. The early railway engineers were drawn from the labor force trained by still earlier enterprises. Almost all of the first railroad inventors, the Stephensons and Trevithick in particular, were engineers with coal-mining firms — although Evans had a connection with steamboats. The importance of this will be discussed below.

The earliest firms in the business of building locomotives drew on experiences in textile machinery and related production. While such related sources were of tremendous help they were hardly adequate. (A famous test of 1837 "proved" that railroads could not go up hills although they had been doing it for years.) The railroads' needs stimulated technical education and a generation of trained engineers, some of whom were themselves easily drawn off into related areas. Schools set up to train engineers did not disappear when the railroads' needs became less pressing, but merely shifted their expertise and facilities into related educational channels.

These two features of the effect of the railroads, which will be developed more fully in the following section, point up the fact that an innovation can have important impacts on economic and cultural activity quite unrelated to the original impetus, because of the reluctance of resources to return to the *status quo ante* once the original stimulus is removed.

Another not unimportant institutional arrangement created by the railroads' effects that did not wither away was the government regulatory commission. Even before the Interstate Commerce Commission was formed, the intrinsic interstate nature of railway operations had led, in a groping pragmatic way, to federal subsidies to railroad construction. But *ad hoc* as they were, federal subsidies left behind them no established principle to intervene in other industries. They did, however, create an environment in which the subsidized rail network seemed accountable to the national government, an environment which, together with the railroads' "natural monopolies," led to the creation of the regulatory agency.

The development of effective regulation was a long and difficult process full of mistakes. The legal problems that arose, the techniques of hearings, the principles of "fairness" that evolved, the methods of enforcement, all developed gradually, and no matter how well-deservedly they might be criticized, smoothed the path of other regulatory units when Congress established them. This, too, was a technology of sorts, and the importance of a *good* technology can easily be seen by a comparison of the results of well-organized and inefficient regulation.

The Impact on the Progress of Research

Strangely enough, writing and speculation about the motive forces that lie behind innovative activity are surprisingly polarized. One substantial school tends to argue that "necessity is the mother of invention" — that inventive effort is stimulated by need. Another important view is that innovation cannot be directed into "useful" lines of activity — it is the product of restless curious minds that follow aimless, freely associative paths of inquiry. This is not the place to decide definitively between such opposed views. Nevertheless, the history of railroad technology suggests that there is something to both points of view, at least in a modern society.

Railroad inventions had a way of appearing at critical junctures that clearly suggested that they were motivated by "necessity." As was mentioned earlier, the earliest inventors worked for coal mines. The geological processes that created coal deposits frequently left them ensconced in hilly country — a terrain quite unadaptable to canal, or road, or river transportation. In addition, coal is a low-value-per-ton commodity that badly needs cheap transport to make it economically competitive. Finally, we find railroads being gradually invented by people associated with the coal industry at a time when wood fuel was becoming increasingly expensive, thus making the reduction of coal-transport costs an increasingly valuable objective.

Similarly, cheap steel was invented at just the time when heavy traffic on existing railroads was wreaking havoc with iron rails, literally chewing up the rails at a pace that threatened to halt the steady reduction in costs of larger trains and heavier loads. The speed with which the railroads converted from iron to steel rails bespeaks the immediate practical usefulness of this important innovation. And then, as cheap steel made possible the economies of long trains carrying heavy loads at high speeds, a new bottleneck loomed: the difficulty of stopping these huge trainloads. At just about the time this problem was becoming urgent, Westinghouse, a man associated with the industry, developed the air brake — which proved to be one of the most important cost-cutting railroad innovations.

While "need" plays an important role in all of this activity, it does not imply the antithesis of the "aimless curiosity" argument. What happened, in the nineteenth-century, Anglo-Saxon culture, was that the economic-technological problems of industry were exactly those that excited the curiosity of the kind of pragmatic "machinist" who invented things. For example, when trains were colliding with alarming frequency, the advantages to be gained from better brakes were known to all of the mechanically inclined. We live in a society where the sources of curiosity and social esteem are also the fountains of financial regard. In this kind of a world any major effort by society would almost surely turn scientific curiosity and inventive genius into similar lines. In much the same way that railroad building led to the development of capital markets and mass-production techniques, which in turn encouraged other kinds of quite unrelated activity, any substantial expansion of some line of activity tends to stimulate curiosity and innovation in related areas that will structure inventive activity in quite unforeseen ways.

It seems very likely that the large commitment of funds to space research will also have secondary effects in stimulating intellectual curiosity in these areas, and thus tend to attract a commitment of effort beyond that of the main government outlays. On the other hand, conditions have changed since the nineteenth century. Applied research is very costly and it is less likely that anything but relatively basic research could benefit from this secondary effect. Even without large secondary effects, however, it should be clear that research "capacity" is not indefinitely extensible, and large outlays in this field will have restrictive effects on research along other lines. Whether or not this is desirable is hard to determine, but it is worthy of study.

Summary

The central argument of this chapter has been that the pace and economic impact of innovation can only be understood in the context of the entire economy — and the relevant economy is not the one that lies within our national boundaries but the total of the interconnected world economies. Great entrepreneurs and imaginative administrators can alter the pace of development, but as long as our economy operates as it does, based on economic incentives, the scope for changing the pace of development is quite limited.

So large, complex, and competitive is this economy, and so continuous and unremitting is the search for new and better techniques, that it is difficult and rare for an innovation to capture completely some important market in a very short period of time. Basically, the problem is one of competition for scarce resources. Except for the relatively narrow range of flexibility permitted by the slow rate of growth of total resources, the rapid rise of one industry involves the decline of another. If some industry accounts for a significant share of economic activity, it is virtually impossible for an innovation to capture the fraction of annual investment that would be necessary for it to dominate that industry in less than several decades. In the same vein, it is almost impossible for some relatively important industry to alter its fraction of national product by as much as 5 per cent, except over long periods of time, because of the competition of long-established human needs and stable patterns of consumption.[9]

It follows, therefore, that if a space-related innovation, for example, is going to become an important factor in communication or tourism it is likely to take several decades before the transition is complete.[10]

In the case of the railroad we find that the competition of the canal and the turnpike made the initial introduction of the innovation relatively slow and hesitant, and that this hesitance reflected sound economic judgment rather than stodginess and conservatism. We find, further, that as the railroad net was constructed the pace and geographical distribution of investment was very heavily influenced by the world-wide pattern of demand. Rather than viewing nineteenth-century economic development in the United States as a reflection of the impact of the railroad, we find that the pattern of growth of the railways reflected the world and national economies.

On an aggregative basis we find that the impact of the railway was not very large until some time after it passed the stage of an "innovation." Its peak impact, in fraction of total resources utilized,

[9] The only exceptions of which I am aware occurred under pressures of war mobilization and the subsequent demobilization.

[10] I should stress, however, that the impact I am concerned with here is the long-run impact on growth and structure of the economy. The short-run impact on the level of economic activity can be much more noticeable. New capital investment of as little as one half of 1 per cent of national product can make a substantial impact on employment if it is added to an underemployed economy, *provided* that it does not simply replace other investment.

came more than four decades after its introduction, and it was not atypical of other innovations in this regard. In individual domestic industries, however, the impact was more varied, coming very quickly in some — as in the manufacture of steam engines — more slowly in others, such as iron, and slowest of all in coal. These differences in the timing of the impact cannot easily be generalized, depending, as they do, on varying conditions of cost among competing materials, the geography of investment, and the state of foreign competition. This very lack of generality re-emphasizes my major argument: the importance of the economic context for understanding and predicting the impact of innovation. It is a subsidiary argument of this paper that one of the impacts of the railroad on resource use has been its claim on research talents excited to curiosity by the railroad's novelty and stimulated by financial rewards implicit in improving the competitiveness of such a large stock of assets.

At a further remove from generality was the impact of railroad innovation on established institutions. By stimulating the growth of domestic capital markets, foreign investment, centralized production, and urbanization, it affected numerous industries in subtle ways too involved to describe here. By increasing regional economic interdependence the railroad helped both to forge a nation and to lead that nation first to subsidize its growth and then to institute the first government regulatory commission to control it. It is this field, the impact of important innovations on our "style" of life rather than on our level of income, about which we are least able to predict; and it may yet be this field that provides the greatest impact for space innovation.

5

The Railroads: Innovators in Modern Business Administration

Alfred D. Chandler, Jr.
Stephen Salsbury

New Enterprise and the Evolution of Administrative Practices: The Impact of the Railroads, 1829–1860

The American entrance into the space race in 1957 created an enterprise which will in the course of time profoundly influence almost every aspect of life. Measured by any past standards space exploration is a vast undertaking. By 1970, total expenditures may reach $40,000,000,000. The manned moon expedition must conquer many difficult technological problems which require precise data from such diverse fields as biology, psychology, physics, and astronomy. Even in the embryonic stages space administrators must face a host of new and delicate problems such as the relationship of a long-term government program controlling enormous resources in money and scientific talent to the rest of the economy, the initiation and co-ordination of the complex and seemingly unrelated basic research necessary to solve the technological complexities posed by space travel, the relationship of a predominantly professional and highly educated working force to management, and many others not even imagined.

Industries which have radically altered nearly every aspect of national life are not new on the American scene. In fact space is merely the latest in a series of revolutionary enterprises — railroads, automobiles, aircraft — which have transformed the United States from Thomas Jefferson's agrarian society to today's scientific and urban economy. Thus, change and readjustment have become normal. Whole industries have grown to maturity only to fall into decline as new technology renders them obsolete. The natural question about the space effort is, therefore, what impact will it have on the economy?

This chapter is concerned with the evolution of administrative structure and technique in large-scale industries. While it is too early to predict or categorize what innovations the space program will bring, a detailed understanding of changes resulting from past economic revolutions may provide a background that will prove useful in analyzing the space effort's impact upon administrative practice in the years ahead.

Although space and railroads have few technological similarities, they both involve the management of large and complex organizations, and in this sense, nineteenth-century railroads may be comparable to space in the twentieth century. Certain it is that railways forced a sharp break with traditional patterns. In three decades between their introduction in 1829 and the outbreak of the Civil War, railroads set precedents which profoundly affected every aspect of the industrial world. In truth, it is not too much to say that railroads created modern administration — that is, they moved business activity away from organizations run by entrepreneurs with the aid of personal trustees, relatives, and the like to corporations with a systematized, bureaucratic management.[1]

Early Railroad Development

The coming of modern administration can best be understood by focusing on how and why America's first big businesses grew, and in what ways railroads developed methods to handle the problems arising from the size and complexity of their operations. Robert Stephenson's perfection of the steam locomotive on England's Liverpool and Manchester Railway, in 1829, immediately sparked many projects in the United States. These were of two kinds. The first was the grand enterprise meant to carry the products of the Great West to the cities on the Atlantic seaboard. In this category fell the Baltimore & Ohio, the New York and Erie, and the Western Railroad of Massachusetts. The second included the less ambitious projects, which generally ran north and south and connected well-established population centers. Of these, the Camden and Amboy, the Boston and Providence, the Philadelphia, Wilmington and Baltimore, and the Boston and Lowell were typical.

The great east-west roads soon dwarfed any previous enterprises. By the early 1850's four trunk lines connected the Atlantic with the western waters: in 1851, the New York and Erie reached Dunkirk on Lake Erie; in 1853, the Baltimore & Ohio arrived at Wheel-

[1] H. H. Gerth and C. Wright Mills, *From Max Weber: Essays in Sociology* (New York: Oxford University Press, 1958), pp. 196–197.

ing on the Ohio River; the same year the several short lines connecting Albany with Buffalo combined to form the New York Central, and the Pennsylvania's tracks reached Pittsburgh. The 1850's witnessed an enormous railroad boom. By 1855, no less than thirteen companies operated systems that exceeded two hundred miles in length, and by 1860, that number had increased to thirty-one.[2] By the eve of the Civil War, rails had replaced canals as the leading inland transportation method, and in the north an iron network linked most important cities and towns from the Atlantic to the Mississippi River.

Railroads and Other Enterprises Compared

From the first, railroad managers faced unique problems. Measured in strictly financial terms, railroads quickly overshadowed all contemporary factories or other transportation ventures. The Western Railroad in Massachusetts, which in 1842 had but 160 miles of single track, cost more than $7,000,000; by 1854, the Western's capital amounted to more than $10,000,000 although it operated no greater mileage.[3] By contrast, the completed Erie Canal, more than 360 miles long, cost only $7,000,000.[4] In 1860, the New York Central, which paralleled the canal, had invested more than $30,000,000 in property, track, and rolling stock, and this was but the beginning: by 1883 the Central operated 953 miles of line with a total investment of nearly $150,000,000. And the New York Central, which owned four times the mileage that it directly operated, was not the largest system. By the time the Pennsylvania had completed its expansion program between 1869 and 1873 its total investment approached $400,000,000.[5] River steamers required much

[2] For mileage in the years before the Civil War, see Alfred D. Chandler, Jr., *Henry Varnum Poor, Business Editor, Analyst, and Reformer* (Cambridge: Harvard University Press, 1956), pp. 267–268.

[3] *Eighth Annual Report of the Directors of the Western Railroad Corporation to the Stockholders* (Boston: Dutton and Wentworth's Printer, 1843), p. 20; *Twentieth Annual Report of the Directors of the Western Railroad Corporation to the Stockholders* (Springfield: Samuel Bowles and Co. Printers, 1855), pp. 18–19.

[4] Carter Goodrich, *Government Promotion of American Canals and Railroads 1800–1890* (New York: Columbia University Press, 1960), p. 54.

[5] Henry V. Poor, "Railroad share list, including mileage, rolling stock, etc.," *American Railroad Journal*, 29 (January 14, 1856), 24–25; Chandler, *op. cit.*, pp. 145, 320; "Statement of the Mileage, Capital, Operations, etc., of the Railroads of the U.S. for 1883," *Poor's Manual of the Railroads of the United States: 1884; Report of the Investigating Committee of the Pennsylvania Railroad Company Appointed by Resolution of the Stockholders at the Annual Meeting Held March 10, 1874* (Philadelphia: Allen, Lane and Scotts Printing House, 1874), p. 115.

less capital than canals; in fact "the construction cost of a single mile of well-built railroad was enough to pay for a new and fully equipped river steamboat." [6] Of manufacturing concerns, textile factories were the largest, but at mid-century only the biggest mills cost as much as $500,000. In fact, in 1850 only forty-one American plants had a capitalization of $250,000 or more.[7]

Railroads differed from canals and factories in very important respects. In the 1850's even the largest manufacturing concerns confined their operations to one or two specific locations which made it possible for managers to view their entire establishments in an hour or two, or to confer with any employee within a matter of minutes. Canal managers, although they supervised works hundreds of miles long, had limited duties confined mainly to routine maintenance and toll collecting. Independently owned boats performed all actual transportation upon them.

By the 1850's many railroads operated systems hundreds of miles in length, and their managers not only supervised maintenance, but were responsible for all movement upon the line since railroads owned and operated all vehicles run upon them. No manager on the Erie or Pennsylvania railroads could inspect their domain in less than several days, and close supervision of subordinates was impossible. Despite this, railroads demanded operational precision that rivaled that of a factory mass-producing complex machinery with interchangeable parts. Safe operations could be achieved only by a strictly disciplined work force, acting in accordance with the most stringent rules. Disobedience or laxity by engineers or conductors meant possible collision, property destruction, or death.

But equally as vital, railroad managers had to make decisions based on data available to them only with detailed planning and organization. A method had to be devised to keep track of freight cars, put them at places where they were needed, and ensure that they did not accumulate in jams at terminal points or lie idle along the line. Railroads faced accounting problems of unprecedented complexity. Collecting revenues produced situations not even encountered in contemporary banking institutions, for each day large quantities of cash flowed through the hands of dozens, if not hundreds, of far-flung ticket and freight agents, passenger-train conductors, purchasing agents, and other officers.

Accounting for and ensuring the honesty of transactions involv-

[6] Lewis C. Hunter, *Steamboats on the Western Rivers* (Cambridge: Harvard University Press, 1949), p. 308.

[7] Evelyn H. Knowlton, *Pepperell's Progress, History of a Cotton Textile Company, 1844–1945* (Cambridge: Harvard University Press, 1948), p. 132.

ing revenue was the essence of simplicity compared to the other accounting problems. The great capital investment required to construct the systems normally resulted in a financial structure that included a substantial bonded indebtedness and a consequent large fixed cost. On Massachusetts' Western Railroad, for example, fixed costs exceeded operating expenses for several years, even after the line had been completed through to Albany.[8] Moreover, capital expenditures seldom stopped but kept growing as increased traffic necessitated more equipment, larger yards, bigger terminals, and double and quadruple tracking of main lines.

Bondholders and stockholders placed a premium on profitable operation. For management, however, it was not easy to differentiate between the "profitable" and "unprofitable" services. The rapid and continuing growth of the physical plant obscured the division between capital expenditures on one hand, and normal maintenance and depreciation on the other. The determination of what it cost to move each class of goods, a vital factor in shaping decisions about service, involved complex calculations. To assess a railway's ability to bid for hauling bulky seasonal commodities like grain, management had first to separate the passenger expenses from the freight, and then those for the grain (terminals, cars, engines, maintenance, capital, etc.) from those for the other freight. Setting rates soon demanded a careful analysis of both the competition and the limitations imposed by the railroad's internal economic position. And by the 1850's massive interregional freight traffic added the complications of fixing joint rates for goods traveling on two or more carriers. For all this railroad management needed precise information about maintenance costs (depreciation of track, rolling stock, depots) as well as operating costs (wood, oil, water, labor, etc.). The collection of such statistics required careful organization and routine methods for employees to record data and pass information along to the top management.

The Application of Traditional Administrative Practices — The Case of the Western Railroad

Early railroad executives did not immediately recognize that their tasks called for organizational skills far beyond those required in other contemporary businesses. Most early systems were started not as enterprises in themselves, but as means to other, more im-

---

[8] In 1842, for example, fixed costs ran $310,000; operating expenses $266,-619.30; *Eighth Annual Report of the Directors of the Western Railroad Corporation to the Stockholders* (Boston: Dutton and Wentworth Printer, 1843), p. 33.

portant ends. The capitalists who created the large industrial complex at Lowell in the 1820's built the Boston and Lowell Railroad in the 1830's to serve their cotton mills. Significantly, one of their own, Patrick Tracy Jackson, was the dominant man in the corporation. Boston merchants started the Boston and Worcester to expand their city's commercial horizons, and Nathan Hale, editor of the *Boston Daily Advertiser*, headed the project.

It is not surprising, therefore, that managerial advances in the 1830's were slight. Lines seldom exceeded fifty miles in length; this was especially true of the numerous north-south systems which linked centers such as Boston and Providence, or Philadelphia and Wilmington. Even in the 1840's long lines were few, as the Baltimore & Ohio and the New York and Erie inched westward at a slow pace. And slack business on the comparatively long South Carolina and Georgia railroads minimized problems.

The experience of America's first heavily trafficked system, the Western Railroad of Massachusetts, quickly demonstrated the need for fresh approaches. The Western, completed in December of 1841, stretched for nearly a hundred and sixty miles between Worcester and Albany; its single-track mountainous line was the major link in the rail route connecting Boston and Albany, at that time two of the nation's most important commercial centers.

Safety soon became the major problem on the Western where it seemed difficult to prevent derailments and collisions. A short road, like the adjacent forty-four-mile Boston and Worcester had been able to prevent wrecks by timing train movements so that morning runs were over before afternoon trains left their terminals; and passing tracks were limited to one or two locations. Such solutions, which minimized the co-ordination and discipline required of employees, were impossible on the Western. There, for example, the morning passenger from Worcester to Albany started its run at 9:30 A.M.; it did not arrive in Greenbush (west bank of the Hudson opposite Albany) until 6:35 P.M. Initially, the Western scheduled three trains a day each way (two passenger and one freight) between Worcester and Albany. This meant a total of twelve daily meets, or times when trains going in opposite directions passed each other.[9] Even assuming that there were no extra movements or work trains on the line, the scheduling problem on a single-track, unsignaled mountain system was difficult. Even before completion, the Western suffered a series of fatal wrecks culminating in a dis-

[9] See timetable in Massachusetts *Senate Document*, No. 55 (1842), pp. 16–18.

astrous head-on collision of two passenger trains on October 5, 1841. This wreck killed a conductor and a passenger, and injured seventeen others.[10]

Reaction was instantaneous. The Massachusetts legislature launched an investigation into the company's management. The *American Railroad Journal and Mechanic's Magazine* editorialized, "We have some recollection of having seen the rules and regulations for the . . . conduct of trains on the Western Railroad . . . but there must be some fault in the regulations or in the mode of enforcing them. We are," concluded the *Journal*, "certain that such accidents are not necessarily attendant upon the railroad system, and that with proper care and precaution, most, if not all, that has happened might have been avoided." [11]

Thus, problems inherent in the operation of a large railway, rather than theory or previous experience, forced the Western's management to review the system's whole administration. It quickly became apparent that there was general confusion and laxness in the transmission of orders regulating train movements from the top management to the train crews. To remedy this, the Western's board of directors appointed a special committee composed of three directors, Elias Hasket Derby, Nathan Carruth, and Abraham Lowe (two Boston businessmen and a physician) and the corporation's Chief Engineer, George Whistler.[12] The rules recommended by the committee and adopted by the directors instituted the railroad's first attempt to erect a structure specifically designed to meet the new problems resulting from the Western's size and complexity.[13]

Although the new rules focused on safety, they attempted to fix definite responsibility for each phase of the company's business, drawing solid lines of authority and command for the railroad's administration, maintenance, and operation. At the apex of power, just below the president, was the chief engineer whose duties were soon assumed by a new office, that of superintendent. The railroad was divided into three divisions, and over each was placed a roadmaster who had direct responsibility for the repair of track, road-

[10] Manuscript records of the Western Railroad, Clerk's File, 1842, Document No. 2; also Western Railroad MSS, Directors' Minutes II, pp. 244–245 (both documents in Boston and Albany Manuscript Collection, Baker Library, Harvard Business School); and Massachusetts *Senate Document*, No. 55 (1842), p. 5.

[11] *American Railway Journal and Mechanic's Magazine* (September 1841), p. 161.

[12] Western Railroad MSS, Director's Minutes I, pp. 305–306.

[13] The Western's new rules are outlined in the Western Railroad MSS, Clerk's File, 1841, Document No. 104.

bed, and bridges. The company asked that each roadmaster keep a "journal of his operations" and make formal monthly reports to the chief engineer. The new directive created the position of master of transportation, with an office at Springfield, to control all freight and passenger traffic, and to maintain the engines and other rolling stock. His authority along the line was enforced through subordinate masters of transportation for each division. Reporting to the latter officials were the various station agents, who not only sold tickets, and received freight, but procured wood fuel for the locomotives as well. The company's major shops at Springfield were supervised by the master mechanic, who in turn had deputies at various terminals and roundhouses along the line. The strictest rules governed the operating employees (locomotive engineers, conductors, brakemen, etc.) who each had a definite place in a chain of command through which authority flowed downward from the chief engineer through the master of transportation, the deputy masters of transportation for each division, to the conductors who had absolute responsibility for their trains. To prevent collisions, exact timetables were published and placed in the hands of conductors with detailed instructions to follow should breakdowns or other factors delay trains. In addition, no schedule changes could be made without the permission of the chief engineer, and then only after written notice had been received by all concerned employees.

The Trunk Lines of the 1850's Revolutionize Administrative Methods[14]

The Western's management responded spontaneously to the crisis of the moment; and it had only dim awareness that railroads created novel administrative problems. But as large-scale systems multiplied, the picture changed. By the 1850's, three great trunk lines, the Baltimore & Ohio, the New York and Erie, and the Pennsylvania, had managers who not only recognized the special complexities of railroad administration, but who were actively engaged in evolving new structures to meet them. The Erie's General Superintendent, Daniel C. McCallum, saw clearly that the methods that had worked on the small systems were inadequate for the large ones: "A Superintendent of a road fifty miles in length can give its business his personal attention and may be constantly on the

[14] Much of the following has appeared in a more detailed form in Alfred D. Chandler, Jr., "The Railroads: Pioneers in Modern Corporate Management," *Business History Review*, 39 (Spring, 1965), 17–40.

line engaged in the direction of its details." Under such circumstances, he reasoned, any managerial method "however imperfect . . . may prove comparatively successful." In running a railroad five hundred miles long, however, McCallum maintained "any system which might be applicable to the business and extent of a short road would be found entirely inadequate. . . . I am fully convinced," he concluded, "that in the want of a system perfect in its details, properly adapted and vigilantly enforced, lies the true secret of their [the large system's] failure; and that . . . [the] disparity of cost per mile in operating long and short roads, is not produced by a difference in length, but is in proportion to the perfection of the system adopted." [15]

McCallum represented a newly emerging group of professional managers who, as civil engineers, had built the great east-west lines, and then had turned to administering the great enterprises which they had constructed. Among the most prominent of these were Benjamin H. Latrobe, Chief Engineer of the Baltimore & Ohio, J. Edgar Thomson, the Pennsylvania's President, John B. Jervis, the Michigan Central's builder and Chief Engineer, and George B. McClellan, the Chief Engineer of the Illinois Central. Latrobe, Thomson, and Jervis had worked up the professional ladders from the late 1820's and were among the very first of the new type of professional engineer.[16] Significantly, only one, McClellan, had any connection at all with military life and he was the least innovative of the lot. These men did not borrow; they approached their brand-new problems of building an administrative structure in much the same rational and analytical way as they approached that of building a railroad or a bridge.

The Baltimore & Ohio Adopts a Functionally Departmentalized Administrative Structure

The Baltimore & Ohio, oldest of the trunk lines, was the first to systematize its operations. Its President, Louis McLane, and its Chief Engineer, Benjamin H. Latrobe, acted not from panic — as did the Western's managers — but from the recognition that as the

[15] *Reports of the President and the Superintendent of the New York and Erie Railroad to the Stockholders for the Year Ending September 30, 1855* (New York, n. d.), p. 34.
[16] For Jervis, Latrobe, McCallum, and McClellan see Dumas Malone, ed., *Dictionary of American Biography* (New York: Charles Scribner's Sons, 1946), Vol. X, pp. 59–61; Vol. XI, pp. 25–26, 565, 581–582; for Thomson see William B. Wilson, *History of the Pennsylvania Railroad Company* (Philadelphia, 1899), Vol. II, pp. 238–239.

railroad pushed westward across the Allegheny Mountains to Wheeling it would need a "new system of management."[17] Up until the new system was adopted in 1847, the corporation operated under plans formulated in 1834, when it had completed only eighty miles of track between Baltimore and Harper's Ferry.[18] The old rules did little more than outline in scant detail the duties of the principal officers; they made no attempt to create a formal working organization, and left unclear the relationship between the various officers.

The Baltimore & Ohio's new managerial plan set forth in a manual, *Organization of the Service of the Baltimore & Ohio Rail-Road,* contained one basic innovation: it departmentalized the road's functions into two separate spheres, finance, and operations. Over-all fiscal responsibility centered in the company's treasurer, who not only reviewed the internal transactions, but also handled external financing, including the routine arrangements for assigning shares of stock or bonds to merchants or bankers who had agreed to market them, assured the proper recording of the sale or other transfer of securities from one person to another, and sent out dividends and interest payments. Directly subordinate to the treasurer was the secretary, whose duties later were taken over by an official entitled comptroller. This officer was wholly concerned with internal transactions; he inspected all passenger and freight accounts and exercised supervision of those who routinely handled the company's monies. Beneath the secretary was the chief clerk, into whose office at the corporation's Baltimore headquarters flowed required receipts and reports from all agents and conductors along the system who received or disbursed funds. The chief clerk's office compiled and checked this information, and issued "daily comparisons of the work done by the road and its earnings with the monies received therefore."[19] Daily figures were in turn summarized into monthly reports. Thus was made available the data so vital to decision making by top management, and for checking upon the honesty and efficiency of the employees. In 1847, however, there was still little attempt to break down operating expenses into their component parts or to allocate costs against the type of goods and passengers carried.

[17] *Organization of the Service of the Baltimore & Ohio Railroad, under the Proposed New System of Management* (Baltimore, 1847), p. 3. The new system was accepted by the board on February 10, 1847.

[18] *Laws and Ordinances Relating to the Baltimore & Ohio Railroad* (Baltimore, 1834).

[19] *Organization of the Service of the Baltimore & Ohio Railroad,* 1847.

The new operating department was placed under the control of a professional engineer who had the title of general superintendent. He supervised three distinct subdepartments: maintenance of way under the master of the road, machinery (maintenance of rolling stock) under the master of machinery, and transportation under the master of transportation. Of the three subofficials the master of transportation was the most important. Everything concerned with the forwarding of passengers and freight was his responsibility, and it was his duty to employ "with the concurrence of the general superintendent and president, all officers and hands necessary" [20] for his department. This included engineers, firemen, conductors, fuel and lumber agents, and depot agents. The latter supervised six types of employees — clerks, weight masters, car regulators, laborers, watchmen, and porters.

The general superintendent was the road's key administrator. Except for the revenues, his office was the central focus of both authority and communication, and into it flowed a series of reports. Each of the operating department heads forwarded their weekly and monthly results. The master of machinery, for example, was to report on "the conditions and performance during the week of each locomotive and engine in service or under repair — and the condition of the cars, as also of the stationary machinery and workshops — and will present a monthly estimate of the probable expense of their repair during the ensuing month." [21] Besides reading reports, the senior operating executive constantly reviewed progress with department heads, inspected the road, and conferred with the president and the road's financial officers.

By the 1847 reorganization, Latrobe and McLane set up one of the very first functionally departmentalized, administrative structures for an American business enterprise. As the departments took over the day-to-day routine operating decisions, the president was able to concentrate more effectively on the long-range activities of raising and allocating funds.[22] While such a departmentalized structure would be expanded and refined as the railroads grew, it remained essentially the organization by which American railways were to be administered.

[20] *Ibid.*
[21] *Ibid.*
[22] Possibly one reason for the change was that President McLane had no time for day-to-day operations. For the two previous years, he had been in England raising money for his road and also acting as United States Minister, in which capacity he helped to negotiate the Oregon Treaty, *Dictionary of American Biography,* Vol. XII, p. 114.

McCallum's Erie Refines the Baltimore & Ohio's Structure

The first refinements to Latrobe's organization came on the New York and Erie, which after it reached the Great Lakes in 1851, was the largest railroad in the United States. However, it suffered from size, and in 1853, after the railroads linking Albany and Buffalo consolidated into the New York Central, the Erie also faced formidable rail competition for the traffic of the Great West. Alarmed because the Erie's costs per mile were higher than those on the shorter roads, the company's board desired a system that would ensure a more precise accountability for expenses and a more effective appraisal of men and managers. This the directors hoped to achieve by making available "comparisons of the expenses of the various operations with those of other similar roads, with the several divisions of the road itself; and the expenses of different conductors, engine-men, etc., with each other." [23]

By the end of 1853, the directors had split the Erie into five geographic divisions, each about a hundred miles long, and they had also made the separation between operation and finance in the same manner as Latrobe's. In 1854 they picked Daniel C. McCallum, at the time superintendent of one of the new divisions, to be the general superintendent. McCallum, the inventor of an inflexible truss bridge and an able engineer, approached the Erie's management in much the same way as he designed bridges. His great strength was in sharpening lines of authority and communication, and in stimulating the flow of the minute and accurate information which top management needed for the complex decisions it was increasingly being called upon to make. Hourly, daily, and monthly reports, more detailed than those called for earlier on the Baltimore & Ohio, provided this essential information.

The hourly reports, primarily operational, gave by telegraph the location of each train and the reasons for any delays or mishaps.[24] The information thus received was tabulated, and proved vital in the elimination of bottlenecks and other trouble spots. McCallum's use of the telegraph impressed other railroad managers, because it demonstrated that wires were more than a means to make trains safe, but were also a device to improve co-ordination and better administration.

[23] *Report of the Directors of the New York and Erie Railroad Company to the Stockholders, November 1853* (2nd ed.; New York, 1853), pp. 47–48.
[24] *Reports of the President and the Superintendent of the New York and Erie Railroad, 1855.* McCallum gives a full account of his reporting systems in this report, pp. 34–35, 51–54.

Daily reports, the real basis of the system, were required from both conductors and station agents. They covered all important matters of train operation, and the handling of freight and passengers. These reports provided information from two different sources on train movements, car loadings, damages, and misdirected freight, and acted as a valuable check on the honesty and efficiency of both conductors and agents. Engineers, too, were required to make daily reports. These were consolidated into monthly statements, giving for each engine the miles run, the operating expenses, the cost of repairs, and the work done. Such data, flowing up through the superintendent's office of each geographic division to McCallum, made it possible for him to make comparative appraisals between the different divisions, and between them and other roads. In addition, the success of experimental motive power could be easily evaluated.

Besides assisting in operations, these statistical data were essential in rate making, for only analysis of these reports could provide the information necessary to determine what were the costs of carrying an item, and whether, therefore, the charges produced a profit or not. McCallum also realized that rates depended on more than costs. The Erie had lost money because it had raised rates which it had found "unremunerative" only to discover that in so doing they had threatened to "destroy this business." [25] Higher rates by reducing traffic had cut net revenues. To guard against such a result required "an accurate knowledge of the cost of transport of the various products, for both long and short distances." Important too was knowing which way the item was moving along the line, for prices should be "fixed with reference to securing as far as possible, such a balance of traffic in both directions as to reduce the proportion of 'dead weight' carried." Unused or excess capacity on a return trip warranted lowering prices for goods going that way.

McCallum's innovations received widespread attention. Henry Varnum Poor, editor of the *American Railroad Journal,* credited McCallum with increasing efficiency while reducing the working force. The New York State Railroad commissioners described the Erie's new managerial system in its annual report. Even a popular magazine like the *Atlantic Monthly* devoted an article to the subject in 1858. In England, Douglas Galton, a leading British railroad authority, described the Erie's management in an 1857 Parliamentary report. Unquestionably McCallum's principles and procedures

[25] This and the following two quotations are from *ibid.,* p. 79.

had a significant impact on the development of the internal organization of the large business enterprise.[26]

The Pennsylvania Adopts a Decentralized, Divisional Structure with Line and Staff Officers

The Pennsylvania, rather than the Erie, tested and further rationalized McCallum's concepts of large-scale administration. Before 1860, the Erie fell into the hands of unscrupulous financiers, who like its notorious Treasurer, Daniel Drew, cared little about efficient administration. McCallum soon retired, and developed a profitable bridge-building business. On the Pennsylvania, however, engineers rather than speculators continued to run the road. J. Edgar Thomson, the builder and first operator of the Georgia Railroad, came to the Pennsylvania in 1849. In 1852 he became its president, and he continued to control its destinies until his death in 1874.

In 1857, Thomson rebuilt the Pennsylvania's managerial structure in the image of McCallum's Erie. Thomson divided his system into several geographic divisions, each managed by a superintendent who reported to a general superintendent with over-all operational responsibility for the system. His organization established solid and clear lines of authority, and ensured a steady flow of data to the top management.

The Pennsylvania's main achievement was a clarification in the relationship between the central office, and the geographic subdivisions. Both the central headquarters and the several divisions carried on at least three functional activities — transportation, maintenance of way, and maintenance of locomotives, rolling stock, and other machinery. Thomson explicitly delegated the full powers to control the road to the officers in charge of transportation — to the general superintendent in the central headquarters and to the division superintendents in the geographic subunits. The other functional officers at the headquarters set standards and procedures but could not order their subordinates in the divisions when to work and what to work on. On the Pennsylvania, therefore, the division superintendent directed the daily work of all men in his division. This meant, for example, that all the workers in the division's shops were under his control; while the master of machinery set rules and standards for "the discipline and economy of conducting

[26] Chandler, *Poor*, pp. 147–148, 153; *American Railroad Journal*, 29 (May 3, 1856), 280.

the business of their [the division superintendent's] shops."²⁷ In short, this was the beginning in industry of the line and staff system where the executives on the line of authority handled the people, and the other officers, the staff executives, handled things.

The decentralized, divisional railroad structure, with line and staff officers, that emerged on the Pennsylvania, and which became characteristic of such major systems as the Michigan Central, Illinois Central, and the Chicago, Burlington & Quincy, was neither natural nor inevitable. British railroads used a centralized "departmental" type of organization where the general superintendent did not delegate his authority to the division superintendents. Instead, each functional officer on a regional division reported directly to and received his orders directly from his functional superiors in the central office.

Centralized Administration Evolves on the New York Central

In the United States, the New York Central developed a centralized departmental structure. Its history differed radically from the other three great trunk lines. Unlike the Pennsylvania, the Baltimore & Ohio, or the New York and Erie, the Central was not built as a single grand enterprise, nor did its construction produce a group of able, professional engineer-managers. Instead, the Central was created from ten separate short lines by financiers and politicians.²⁸ The system's first senior executives, Erastus Corning, Dean Richmond, John V. L. Pruyn, and Edwin D. Worcester, were among New York's richest men, and they were powerful leaders in her Democratic Party. There was not a professional engineer among them; even the new General Superintendent, Chauncy Vibbard, had no training or apprenticeship comparable to that of the operating managers of other trunk lines.

The New York Central was one of the very first great consolidations of a number of different incorporated enterprises. Its major problems were financial and legal, and its early managers focused on these, paying relatively little attention to the development of a

²⁷ This quotation is from *Organization of the Pennsylvania Railroad, 1857*, p. 7. In 1857 there were only two resident engineers, each reporting to the general superintendent. After 1863 there was a resident engineer for each of the three divisions, *By-Laws and Organization for Conducting the Business of "The Pennsylvania Railroad Company," as revised and approved by the Board of Directors, May 13, 1863* (Philadelphia, 1863), p. 14.

²⁸ The carrying out of the consolidation is described in detail in Frank W. Stevens, *The Beginnings of the New York Central Railroad* (New York: G. P. Putnam's Sons, 1926), Chapter 17.

rational operating structure. Chauncy Vibbard, in addition to his duties as general superintendent, continued to run a profitable liquor business in New York City, and in 1861 he was elected to Congress.[29]

Vibbard did not systematize the road's structure. Although the Central had five regional divisions, each headed by an assistant or "deputy" superintendent, Vibbard delegated little authority. As late as the 1850's, he still made verbal arrangements to buy fuel wood.[30] Vibbard's many outside interests prevented him from direct supervision of all activities, and eventually strong autonomous functional departments grew up at the road's main headquarters. Under this system the division superintendent's duties ended with control of train movements; and those responsible for maintenance of way and bridges continued to report to the chief engineer, the shops to a master of machinery, or a master of car repairs, and the station and freight agents to the general passenger or the freight agent. Thus, the officers at the headquarters, who on the Pennsylvania held staff advisory positions, on the Central had direct line control of and responsibility for the actions of workmen at the division level. Vibbard's structure, always haphazard, and never thought out, became formalized into an explicitly centralized departmental type when Cornelius Vanderbilt and his son William took control of the line after the Civil War. Although the Central's management structure remained different from those of other large systems, the demands of running a large railroad forced the Vanderbilts to formalize the structure and sharpen lines of authority and responsibility.

In three short decades, railroads transformed American business organization. The highly technical and complex requirements of railroad construction and operation developed a professional managerial class sharply different from the original entrepreneurs — local businessmen, bankers, or community leaders — who personally shaped and supervised small concerns. On large railroads the old managerial patterns simply would not do, and the new executives soon recognized this and set about formally to create administrative structures specifically designed to the needs of large-scale, geographically scattered, and technologically complex enterprises. The decentralized divisional structure with line and staff officers which finally emerged on the Pennsylvania was not inevitable. But

---

[29] *Dictionary of American Biography*, Vol. XIX, p. 263.
[30] Alvin F. Harlow, *The Road of the Century* (New York: Creative Age Press, 1947), p. 96.

the great distances on American railroads contrasted sharply with the shorter systems in Europe and in Great Britain, and thus made the centralized "departmental" type of organization, typical of English roads, less practical, as the New York Central found out. The moral of this story is that administrative problems are by no means automatically solved. Nevertheless, railroads like the Pennsylvania created the managerial skills and bureaucratic structures which, with slight modifications, made possible the administration of the new industries that in the period between the Civil War and the First World War did so much to change America from an agricultural and trading economy to a vast industrial and urban society.

The American Railroads, 1860–1900: The Growth and Management of the Nation's First Private Bureaucracies

The decade following 1865 brought enormous changes to America's railroad network. Before that time major attention focused on problems associated with building the first lines: financing, technological improvement, and operation. Prior to the Civil War, top management built an administrative structure capable of directing systems five hundred miles long.

The 1850's saw concrete recognition that railroads differed from other contemporary enterprises and that they demanded new administrative structures to meet unique problems. The line and staff organization which evolved on the Pennsylvania enabled management to increase safety, efficiency, and economy. Data flowing from each division to the general superintendent's office made possible more exact scheduling, maximum use of rolling stock, evaluation of new equipment, and provided an accurate basis for determining operating costs.

Competition Among the Trunk Lines

Operational matters and construction continued to concern management after the War, but they were dwarfed by new troubles. The four great trunk systems, the Erie, the Pennsylvania, the New York Central, and the Baltimore & Ohio, entered the post-Civil War decade with compact five-hundred-mile lines that connected Atlantic port cities to terminals on the waters west of the Allegheny watershed. None reached beyond Pittsburgh, or western New York State; and all depended upon newly constructed connecting lines for the traffic which came to them from the West. From Pittsburgh, the Pennsylvania Railroad's western terminus, Chicago, was reached over the northern route of the Pittsburgh, Fort Wayne and Chicago, or a more southerly combination of railroads that included the Panhandle (from Pittsburgh to Columbus) and the Columbus, Chicago and Indiana Central (from Columbus to Chicago).[31] A series of short lines linked Buffalo and Chicago. They paralleled the southern lake shore, connecting Erie with Cleveland, Cleveland with Toledo, and Toledo with Chicago. The Atlantic and Great Western connected western New York with Dayton, Ohio. Other lines linked Buffalo with Chicago through southern Ontario and central Michigan.

Several factors profoundly affected the railroad network that stretched between New York and the Middle West. First, although no single corporation controlled trackage from the Atlantic seaboard to Chicago, a heavy freight volume did flow between these terminal points. This was the period when large quantities of grain and livestock moved eastward from Illinois, Wisconsin, Iowa, and Minnesota to Atlantic and overseas markets. And after 1865, oil shipments between the producing regions of western Pennsylvania and the East became significant. In short, through traffic between the Mississippi Valley and the coast was vitally important to the four great trunk lines, the Erie, the New York Central, the Pennsylvania, and the Baltimore & Ohio. Second, because of numerous interconnections, the western lines were not bound to a single eastern trunk but could deliver goods to any or all of the great eastern roads. Third, most of the railroads had substantial fixed costs; this was especially true of the vast majority of systems outside New England, which were financed by bond issues that required large annual interest payments — which accrued regardless of the amount of traffic carried. Heavy fixed charges put a premium on high utilization, for it was only through steady and rising business that the enormous investments could be made profitable.

Control of Competition Through Alliances

From the first days railroad managements had attempted to curb competition and rate wars through alliances both with potential competitors and with connecting railroads. As early as 1854, the Erie, the New York Central, and the Pennsylvania, together with four western railroads and several Great Lakes steamship lines,

---

[31] Julius Grodinsky, *Jay Gould, His Business Career 1867–1892* (Philadelphia: University of Pennsylvania Press, 1957), pp. 56–58.

met in Buffalo and set rates between New York and all places on and west of Lake Erie.³² But these voluntary agreements were abandoned almost as soon as they were made. Weaker lines, suffering the pressures of high fixed charges, inevitably succumbed to the lure of increasing business by lowering rates.

Alliances between connecting roads proved more enduring. The Pennsylvania was the first major system to embark on such a policy. To ensure itself a voice in the management of its western connections, it began to support certain lines financially through the purchase of their stocks or bonds. In 1853, the Pennsylvania legislature passed an act permitting the railroad to "subscribe capital, or guarantee bonds of other companies, to the extent of 15 per cent of its paid up capital." ³³ This act, one of the earliest laws allowing a company to hold the stock of another doing the same business, helped introduce the holding company, the legal device which became so essential to the growth of the large American business enterprise. In 1858, the Pennsylvania's President, J. Edgar Thomson, explained that his company's "policy . . . [of aiding] in the construction of Western Railways designed to facilitate trade to and from . . . [the] road" had compelled an investment of over $1,600,000 in the Pittsburgh, Fort Wayne and Chicago, the Steubenville and Indiana, and the Marietta and Cincinnati railroad companies.³⁴ Further to ensure a connection between Philadelphia and New York, the Pennsylvania signed a treaty of alliance with the "Joint Companies" in New Jersey. This agreement set rates and provided for common use of facilities.³⁵ By 1868, therefore, the Pennsylvania had allied itself with corporations whose tracks touched both the Atlantic and Lake Michigan. The Baltimore & Ohio followed the Pennsylvania's example, making alliances with or aiding

[32] *Report of the Directors of the New York and Erie Railroad Company to the Stockholders, November 1853*, p. 53; also *Eighth Annual Report of the Directors of the Pennsylvania Railroad Company to the Stockholders, February 5, 1855* (Philadelphia, 1855), p. 13.

[33] Henry V. Poor, *History of the Railroads and Canals of the United States* (New York: J. H. Schultz, 1860), p. 471. This act was passed March 23, 1853.

[34] *Eleventh Annual Report of the Pennsylvania Railroad Company to the Stockholders, February 1, 1858*, p. 14; Poor, *History of the Railroads*, pp. 471, 474; *Sixth Annual Report of the Pennsylvania Railroad Company to the Stockholders, February 7, 1853*, pp. 21–26; *Seventh Annual Report of the Pennsylvania Railroad Company to the Stockholders, February 6, 1854*, pp. 6–7, 18–20.

[35] George H. Burgess and Miles C. Kennedy, *Centennial History of the Pennsylvania Railroad* (Philadelphia: Pennsylvania Railroad Co., 1949), pp. 236–237.

connecting roads that ran into Parkersburg, West Virginia, and Columbus, Ohio.[36]

Although alliances with connecting roads proved more satisfactory than those with competing roads, such tactics still did not ensure stability. The Pennsylvania and the Baltimore & Ohio remained in the position of minority stockholders in their western connections; and, as events were to prove, financial strains on the western lines could easily snap alliances and raise the threat of working agreements with, or control by, hostile forces.

Gould Challenges the Alliance System

Competition between the four eastern trunk lines had always existed, but the Civil War's traffic boom minimized problems; after 1867, however, the picture began to change. The Erie Railroad, a developmental line built through sparsely settled mountainous terrain from New York City to Dunkirk on Lake Erie, had suffered financial woes from its inception. Because of its weakness, it soon fell into the hands of Wall Street speculators. In 1867, after a spectacular stock-market war, Jay Gould emerged as the Erie's leader. Gould found himself with a railroad that had an unfavorable route, extremely high fixed costs, and a need for more capital.

Gould's first step in resuscitating his system was an attempt to control certain railroads operating between western New York and Chicago. Gould reasoned that he could not depend upon the western roads to deliver needed traffic voluntarily, and that only absolute control of connecting lines could assure a large volume of freight flowing from the Midwest to New York via the Erie. Thus, Gould leased the Atlantic and Great Western, which was in difficult financial straits because of the failure of its British backer in 1866. Next, he started buying stock in the Michigan Southern, and in the Toledo, Wabash and Western, and negotiating for control of the Indiana Central, and the Pittsburgh, Fort Wayne and Chicago. By these moves he hoped to extend his influence over lines that connected the Great Western with Chicago, and to reach out as far as possible toward St. Louis. The Indiana Central and the Fort Wayne were important freight sources for the Pennsylvania Railroad which quite naturally did not want them to pass into the grasp of the Erie.

Gould's second step was to reduce rates. Throughout most of his

[36] Poor, *History of the Railroads*, pp. 580, 582; Edward Hungerford, *The Story of the Baltimore & Ohio Railroad, 1827–1927* (New York: G. P. Putnam's Sons, 1928), Vol. II, pp. 68, 110–111.

career he tried to attract traffic by lowering charges. He opposed limiting competition to service alone, and when he made rate agreements, he did so "only to break them." [37]

Jay Gould's attempts to seize the Indiana Central, the Pittsburgh, Fort Wayne and Chicago, and the Michigan Southern forced the Pennsylvania's J. Edgar Thomson, and the New York Central's Cornelius Vanderbilt to adopt defensive strategies. They had two clear alternatives: they could continue their attempts to protect themselves through interroad alliances, co-operation, and the negotiation of increasingly formal agreements, or they could expand their roads into large self-contained systems connecting the nation's great economic regions. Either direction involved managerial problems of a significantly different nature than those of the pre-Civil War decades.

The New York Central continued to rely on co-operation and alliances to ameliorate cutthroat competition. Cornelius Vanderbilt and his son William, who inherited his father's empire upon the Commodore's death in 1877, opposed the extension of the Central's tracks beyond New York State. Although Gould's attempt, in 1869, to control the Lake Shore and Michigan Southern, a key link for traffic flowing between Chicago and the Atlantic via the New York Central, caused the Commodore to buy a majority interest in the Lake Shore Railroad, Vanderbilt did not merge his new acquisition into the New York Central.[38] Instead, he placed the line under the separate administration of his son-in-law, Horace Clark, who also used Vanderbilt funds to purchase stock in such systems as the Cincinnati, Hamilton and Dayton, the Ohio and Mississippi, and the Michigan Central. With the exception of the Lake Shore, however, the Vanderbilts remained influential but minority shareholders in the other systems. But their stock speculations made sense strategically, for they were in corporations which provided potential allies on the routes to Cincinnati and St. Louis or in the companies which were in direct competition with the Central or the Lake Shore.

The Vanderbilts Encourage Federation

The Vanderbilts had great faith in railroad federations which would eliminate competition by setting rates and allocating traffic. They looked backward toward the original attempt at co-operation between the trunk lines in the 1850's. The Civil War's traffic boom

[37] Grodinsky, *op. cit.*, p. 596.
[38] *Ibid.*, p. 65.

temporarily ended intense rivalry, which did not re-emerge until sometime after 1865. Management reacted to renewed competition, both in the East and the West, by creating pools. One of the best known of these arrangements, the Iowa Pool, was formed by the Rock Island, and the Chicago and Northwestern, and the Burlington. By this unsigned, informal arrangement each company kept 45 per cent of its passenger receipts and 50 per cent of its freight revenues to cover operating expenses, and paid the balance into a pool which was divided equally among the three roads.[39] Such alliances generally failed, because they were difficult to modify or adjust when conditions changed, and because of the absence of enforcement devices.

Initial failures did not discourage further co-operative attempts. The panic of 1873 reduced business activity, decreased freight, and increased pressure to cut rates. That year the presidents of the New York Central, the Erie, and the Pennsylvania proposed that an association should be created to set rates, but the Baltimore & Ohio's President, John W. Garrett, who was building his own line into Chicago, blocked this by refusing to participate.[40] Competition worsened, and in 1875 Canada's Grand Trunk joined the four original eastern trunk lines as a contender for the midwestern traffic. It created a through east-west route by allying itself with the Michigan Central and the Vermont Central, and immediately cut grain rates.[41] In 1876 the situation was further complicated when the merchants of the various port cities, especially Baltimore, Philadelphia, New York, and Boston, each demanded that the railroads should set special low rates that would aid local merchants in competition for western business with those from rival ports.[42]

In 1877 the embattled railroads took steps to bring order out of chaos. In April they signed a seaboard differential agreement that gave Philadelphia and Baltimore lower prices on western traffic than it did New York and Boston. In July the roads formed the

---

[39] Julius Grodinsky, *The Iowa Pool: A Study in Railroad Competition, 1870–1884* (Chicago: University of Chicago Press, 1950), p. 17.

[40] *Twenty-Eighth Annual Report of the Pennsylvania Railroad Company to the Stockholders, March 9, 1875*, pp. 41–42; Joseph Nimmo, *Community of Interests, Method of Regulating Railroad Traffic in Its Historic Aspect* (Washington, 1901), p. 16.

[41] Edward Chase Kirkland, *Men, Cities, and Transportation* (Cambridge: Harvard University Press, 1948), Vol. I, pp. 498–500.

[42] *Ibid.*, Vol. I, pp. 508–509; D. T. Gilchrist, "Albert Fink and the Pooling System," *Business History Review*, 34 (Spring, 1960), 34; *Thirty-First Annual Report of the Pennsylvania Railroad Company to the Stockholders, March 24, 1878*, pp. 69–70.

Eastern Trunk Line Association, and for its commissioner they selected Albert Fink, who had pioneered in a federation of southern lines that was started in 1875.

Fink moved quickly to set up regional committees of competing roads to meet at regularly specified times to determine local as well as interregional freight rates and classifications.[43] Simultaneously he created a large staff in New York which collected information on existing rates and traffic movements for the use of the committees in their deliberations. Fink then decided that the connecting western and New England lines must be included within his organization. In the summer of 1878, the midwestern roads formed a Western Executive Committee to set rates for and to allocate eastbound traffic. Then, in an agreement signed in December of 1878, the many roads designated a Joint Executive Committee, which would give the final approval of all rates worked out by the regional subcommittees or associations in the East and West.[44] Fink became chairman of this committee, and all cases not decided unanimously were referred to the chairman who made the final award. As a check upon Fink, a board of arbiters was created, which would be called in if a railroad refused to accept his decisions. The board of three included some of the most able and respected railroad experts of the day: Charles Francis Adams, Chairman of the Massachusetts Railroad Commission; David A. Wells, the economist; and J. A. Wright.

By the end of 1878, Fink headed a railroad federation which contained most systems north of the Ohio and east of the Mississippi. But even this organization failed. In Fink's words, "the only bond which . . . held the government together . . . was the intelligence and good faith of the parties composing it."[45] In the early 1880's, Jay Gould, who had left the Erie for greener speculative pastures beyond the Mississippi, again invaded the East and gained control over the Wabash, the Lackawanna, the Central of New Jersey, and the Boston, New York and Erie. Gould and the leaders of other weak lines soon made a shambles of Fink's organization.

By 1881 Gould's successes and Fink's failures finally caused

[43] Gilchrist, "Albert Fink and the Pooling System," *op. cit.*, p. 35.
[44] *Ibid.*, pp. 36–37; *Testimony of Albert Fink [before] United States Senate Committee on Labor and Education, New York, September 17, 1883*, pp. 4–5.
[45] *The Railroad Problem and Its Solution: Argument of Albert Fink before the Committee on Commerce of the U.S. House of Representatives, in Opposition to the Bill to Regulate Interstate Commerce, January 14, 15, and 16, 1880* (New York, 1882), p. 2.

William Vanderbilt to change his strategy. He decided that the Eastern Trunk Line Association could not adequately protect his interests. Although Vanderbilt continued to support the association, he shifted his primary energies to the course of action which the Pennsylvania had adopted to counter Gould's aggressive tactics back in 1867.

The Pennsylvania Prefers Consolidation

The Pennsylvania's management had not opposed interrailroad co-operation, but it felt that Gould could best be checked by a grand plan of expansion that would allow the Pennsylvania on its own track to "reach all important points in the West." [46] Gould's try in 1869 to control the Indiana Central and the Pittsburgh, Fort Wayne and Chicago failed. The Pennsylvania used its superior financial power to lease the Fort Wayne and the Indiana Central. By 1873, the Pennsylvania had leased or purchased full control of a vast integrated network joining most major cities of the Midwest and the Atlantic seaboard, and the line which in 1869 had operated a main track between Philadelphia and Pittsburgh became a vast system connecting New York City with such distant points as Chicago, St. Louis, Cairo, Indianapolis, Cleveland, and Upper Michigan.[47] By such means the Pennsylvania hoped to control traffic from its point of origin to its destination, and to free itself from dependence upon other companies. For a while it went even further. In 1871 it organized the American Steamship Company to operate between Philadelphia and Liverpool.[48] This was to lessen the railroad's reliance upon the port of New York. Later, it moved into oil refining and coal mining to protect its position as a carrier of those commodities.

The Baltimore & Ohio quickly followed the Pennsylvania's lead, and by 1874 it controlled tracts reaching from Baltimore to Pitts-

---

[46] *Report of the Investigating Committee of the Pennsylvania Railroad Company Appointed by Resolution of the Stockholders at the Annual Meeting Held March 10, 1874* (Philadelphia: Allen, Lane and Scotts Printing House, 1874), p. 45.

[47] *Ibid.*, pp. 45–61; *Twenty-Third Annual Report of the Pennsylvania Railroad Company to the Stockholders, February 15, 1870*, pp. 15–20; *Twenty-Fourth Annual Report of the Pennsylvania Railroad Company to the Stockholders, February 21, 1871*, pp. 17–27; *Twenty-Fifth Annual Report of the Pennsylvania Railroad Company to the Stockholders, February 20, 1872*, pp. 14–20; and Burgess and Kennedy, *op. cit.*, pp. 195–240.

[48] *Twenty-Fifth Annual Report of the Pennsylvania Railroad Company to the Stockholders, February 20, 1872*, pp. 27–28.

burgh, Cincinnati, and Chicago.⁴⁹ Gould, as was said, scuttled the Erie in the early 1870's, and decided to try his hand at building a railroad empire west of the Mississippi. But Gould's temporary departure did not stop the drive toward giant self-contained systems. By 1880, the Erie had itself reached Chicago. Vanderbilt's decision, in 1881, to expand the Central merely created another great eastern system.

Meanwhile, Gould's western adventures forced railroads beyond the Mississippi to adopt tactics similar to those that had built the major eastern lines. By the late 1880's vast integrated systems linked most western cities. Among the larger railroads were the Chicago, Burlington & Quincy; the Chicago, Rock Island and Pacific; the Chicago and Northwestern; the Chicago, Milwaukee and St. Paul; and the Santa Fe. All, like the Pennsylvania, were consolidations of numerous smaller roads.

The Impact of Competition on Administration

Railroad competitive strategy profoundly shaped administration. In 1850 operating men with engineering backgrounds — the Erie's Daniel McCallum, the Baltimore & Ohio's Benjamin H. Latrobe, and the Pennsylvania's J. Edgar Thomson — created administrative patterns to run their newly built enterprises. Their structures primarily ensured safe operation and a constant flow of vital data to top management, and their ideas came to be widely accepted throughout the industry. The engineer-managers either ran their systems directly or worked closely with the promoters, usually on-line merchants and lawyers who had nominal managerial responsibilities as presidents and directors.

After 1865 the picture changed. Major entrepreneurial decisions no longer concerned promotion but involved the protection of large capital investments. Nor were top management's problems primarily operational, for the competitive wars which forced alliances and then the growth of self-contained systems caused energies to be concentrated upon the problems of rate making, interrailroad agreements, and consolidation. This necessitated large amounts of money. On the Pennsylvania alone capital increased from nearly $46,000,000 in 1865, to more than $400,000,000 in 1873.⁵⁰ And the

---

⁴⁹ Henry V. Poor, *Manual of the Railroads of the United States for 1870–1871* (New York: H. V. and H. W. Poor, 1870), p. 169; Edward Hungerford, *op. cit.*, Vol. II, pp. 68, 106–108, 155, 220–227.

⁵⁰ Henry V. Poor, *Manual of the Railroads of the United States, for 1868–1869* (New York: H. V. and H. W. Poor, 1868), p. 230.

vast interstate networks also demanded legal and administrative reorganizations to make possible effective control by a single management. Thus, after 1864, the engineer-manager and the part-time merchant-promoter gave way to strong financial leaders familiar with Wall Street money markets. Typical of the new men were the Baltimore & Ohio's John Garrett, the Michigan Central's James Joy, the Illinois Central's William Osborne, the Rock Island's John F. Tracy, the Chicago and Northwestern's Albert Keep, the New York Central's Cornelius and William Vanderbilt, and the Erie's Jay Gould, none of whom were engineers or operating men. Experienced railroad managers like J. Edgar Thomson of the Pennsylvania, Albert Fink of the Louisville & Nashville, and Charles Perkins of the Burlington were the exception.

The Pennsylvania Pioneers in New Legal and Administrative Forms

The holding company, a device which later provided such industrial giants as Standard Oil of New Jersey, United States Steel, and General Motors with a solid legal framework for controlling far-flung interstate industrial organizations, first emerged in full bloom as a result of railroad consolidation. J. Edgar Thomson and Thomas Scott took the lead, and in 1870 the Pennsylvania legislature at their urging created the Pennsylvania Company (as distinct from the Pennsylvania Railroad Company which was the original corporation), which came to hold the leases and stock of those railroads between Pittsburgh, St. Louis, and Chicago which the Pennsylvania Railroad Company came to control. By the end of the Pennsylvania's great period of expansion the legal ownership of the many companies that made up the system resided in the hands of three corporations. The Pennsylvania Railroad Company, the parent corporation, owned and directly administered all lines east of Pittsburgh and Erie. The parent corporation also held the stock of the Pennsylvania Company, which controlled and operated all lines running from Pittsburgh to the northwest and Chicago. And the Pennsylvania Company held the stock of a third major operating company, the Panhandle, which controlled and administered all lines from Pittsburgh southwest to St. Louis. The Pennsylvania's complex legal structure became the standard model for the other great railroad systems that emerged in its wake, and it permanently solved the knotty legal problem of how a corporation chartered in one state could safely and effectively extend into other states.

But legal problems were not the only troubles faced by the Pennsylvania's management. The system had grown from a five-hundred-mile trunk in 1865 to a network of more than five thousand miles. Thomson recognized that a general administrative reorganization was necessary "to secure, by a single management . . . harmonious action throughout the entire system . . . and at the same time, to obtain the best results from the large amounts of rolling stock." [51] Thomson's new organization did not merely bring many formerly separate corporations under a single control retaining the old administrative structures; it built a new administrative form tailored to the Pennsylvania's operational needs.

Thomson created a "decentralized" managerial structure designed to facilitate control at all levels despite the corporation's enormous size. The Pennsylvania was divided into three major administrative units each of which contained an average of about seventeen hundred miles of track. These corresponded to the three great interlocking holding corporations, the Panhandle, the Pennsylvania Company, and the Pennsylvania Railroad Company, which themselves had been shaped with the idea of making them administrative as well as legal units. Each of these regional systems was placed under a "General Manager" who had full responsibility for the "*safe* and *economical* operation of the Roads committed to his charge." He directly controlled the transportation, traffic, and purchasing offices and was responsible "with the assent of the President" for the hiring, firing, and promotion of all executive and administrative personnel within his region.[52] Only the accounting and other financial and legal officers did not report directly to the general managers of the three regions. Those officials were responsible to the main office in Philadelphia.

Each of the great regional systems was further divided into administrative subdivisions. The eastern system, the Pennsylvania Railroad Company, had about five such subdivisions all of which corresponded roughly to what had been independent railroad managements before 1870; but significantly, boundaries were reshaped to meet the needs of traffic flow and administrative oversight. Each of these subdivisions, the longest of which was over five hundred

---

[51] *Twenty-Fifth Annual Report of the Pennsylvania Railroad Company to the Stockholders, February 20, 1872*, p. 16.
[52] *By-Laws and Organization for Conducting the Business of the Pennsylvania Railroad Company to Take Effect June 1, 1873* (Philadelphia, 1873), p. 13; *By-Laws and Organization for Conducting the Business of the Pennsylvania Railroad Company* (Philadelphia, 1881), pp. 10, 22.

miles long, or about the size of the original trunk lines before 1865, was broken into units of about a hundred miles in length.

The greatly enlarged Pennsylvania thus came to have four levels of executives. At the top were the general officers for the railroad as a whole, next the general managers who administered each of the regional systems, then came the general superintendents who handled the regional systems' larger subdivisions, and finally there were the divisional superintendents who managed the small hundred-mile-long divisions.

In effect, the Pennsylvania organization fashioned by its President, J. Edgar Thomson, was the old decentralized, divisional structure of the 1850's writ large. The general officers at the top level made the key entrepreneurial decisions — on over-all rate policies, on expansion, on meeting competition — but they relied on the data which flowed into the head office from the operating divisions. The giant Pennsylvania retained the line and staff feature of the old organization. The general officers set and enforced standards for the operating divisions, whose managers were delegated with the powers to hire, promote, and fire within their own units. The division superintendents, therefore, though their actions were in harmony with the railroad as a whole, had much of the autonomy and authority characteristic of the manager of a small railroad in the 1830's.

Although the Pennsylvania was the first interregional system to rationalize its legal and managerial structure, other railroads carried out comparable reorganizations in the 1880's after they too had expanded through purchase or construction. Among the most important of these were the Chicago, Burlington & Quincy, the Rock Island, the Baltimore & Ohio, and the Southern Pacific.[53]

---

[53] The quickest method to determine whether a railroad had a decentralized structure was to check Henry V. Poor, "List of Officers of Operating Railroads in the United States and Canada, and of the Chief Railroads in Mexico," which first appeared in the 1891 edition of *Poor's Manual of the Railroads*. This gives a full list of executives on all lines and their titles. A road was considered to have a "decentralized" structure when it had at least two units each with their own general manager or superintendent who had a traffic officer directly under him, and if it had no traffic officer in the general office except for a vice president. An article in *The Railway Age*, 10 (November 12, 1885), 710–711, describing and praising the Burlington structure listed the Southern Pacific, the St. Paul, and the Union Pacific as examples of roads consolidating all lines into a single operating unit. But by 1891 the Southern Pacific had been subdivided into two autonomous units — the Pacific System (which included a large, quite autonomous Coast Division) and the Atlantic System; Henry V. Poor, *Manual of the Railroads of the United States for 1891* (New York: H. V. and H. W. Poor, 1891), pp. 916–944, 1365–1369.

While the holding company as a device for legal control of far-flung railroad empires became standard, not all companies emulated the Pennsylvania's managerial method. After all, J. Edgar Thomson was unique. He was an engineer and able operating man who also became a shrewd financier. But in the post-Civil War era financiers dominated, and as a rule they regarded administrative reform as a minor issue.

Among the managers Jay Gould probably represented an extreme. Although from time to time he controlled several groups of railroads that rivaled the Pennsylvania in size, he did nothing to reorganize them into rational administrative units. His control remained personal, and there is little doubt that lack of over-all planning, co-ordination, and appraisal was a major reason why Gould's roads became a "synonym for bad management and poor equipment."[54]

Reorganizing the New York Central

Initially the Vanderbilts did little more than Gould to shape their vast holdings into a consistent framework. When Jay Gould forced Cornelius Vanderbilt into acquiring the Lake Shore and Michigan Southern Railroad in 1869, the Commodore paid no attention to its administration; he merely handed the new company over to his son-in-law, Horace Clark. Not until after Clark's death in 1873, and the consequent disclosure of the road's wretched financial condition, did the Commodore take a stronger hand. The result was a gradual movement toward the Pennsylvania's decentralized divisional structure. The Lake Shore and Michigan Southern and other railroads which the Central later acquired became major units similar to the Pennsylvania's "regional systems." When William Vanderbilt further clarified the outline of his holding, in 1883, each of the great constituent roads had professional presidents who corresponded to the Pennsylvania's general managers. At the main headquarters the executive and finance committee made many of the entrepreneurial decisions and co-ordinated policies between the constituent roads. Significantly, the accounting departments of the operating roads reported directly to the Central's head office.[55] Vanderbilt's structure never attained the highly rational form char-

---

[54] Robert E. Riegel, *The Story of Western Railroads* (New York: The Macmillan Company, 1926), p. 161; Grodinsky, *Jay Gould*, pp. 384, 598–599.

[55] Thomas Cochran, *Railroad Leaders, 1845–1890* (Cambridge: Harvard University Press, 1953), pp. 394–395, 398, 478; *Railroad Gazette*, 17 (December 11, 1885), 785.

acteristic of Thomson's. The Central's component lines were not reformed with a view to developing the most efficient managerial units.

Nevertheless, Vanderbilt's methods of administration through financial control and the allocation of financial resources were particularly significant because they became a model for later empires in mining and heavy industry. J. Pierpont Morgan faithfully attended the New York Central's directors' meetings after 1879, and the financial and general office which directed the affairs of the United States Steel Corporation and other industrial combinations that Morgan helped to create bore a striking resemblance to the Central's head office. While the Pennsylvania's organization had significant similarities to Alfred Sloan's structure for General Motors in the 1920's, there was no historical relationship between these two "decentralized" structures. On the other hand, the New York Central can be said to be the administrative parent of the United States Steel Corporation and many others of the nation's first industrial giants.

Despite the decentralized divisional structure's many advantages, a quite different administrative system finally came to dominate most of America's great railroads. This was the centralized structure of which the Illinois Central was an early example. Here, too, bankers and investors rather than trained railroad men or engineers were in charge.

The Illinois Central Develops a Centralized Management

The Illinois Central's history differed from many other railways: its growth was relatively unhurried, and its financing was sound. Its main line ran north and south, and it was less harassed by Gould's campaigns than the great east-west systems.[56] Even so, the company embarked on an expansion program to protect its business, and by 1888 it operated more than 2,500 miles of track. The Illinois Central's slow growth created no managerial crisis, and, as it expanded, the division superintendents, the superintendents of motive power and maintenance of way, and the traffic officers came to report directly to their department headquarters in Chicago. No one found the time or saw the need to define explicitly the duties

---

[56] J. C. Clarke to Stuyvesant Fish, February 2, 1885, quoted in Cochran, *op. cit.*, p. 299. *Ibid.*, pp. 46–48, has a brief summary of the road's history in these years. More details can be found in Carleton J. Corliss, *Main Line of Mid-America* (New York: Creative Age Press, 1950), pp. 205–255.

of the different officers and the lines of communication between them.

In 1883, the road came under the control of well-established eastern investors, including the Astors, the Belmonts, and Edward H. Harriman. In 1887, a representative of this group, Stuyvesant Fish, became the road's President, and Harriman the Vice President. The two were appalled by the slipshod administrative methods and internal financial procedures, and in 1888 they set about to reform matters. For advice they relied heavily on the Illinois Central's old operating executives. The acting general manager and his subordinates in the transportation department favored moving from a centralized structure to a variation of the Pennsylvania's organization which would break the road into units of management.[57] The traffic manager, on the other hand, recommended that his department should cover the whole system and be completely independent of the transportation executives.[58] He wanted to report directly to the president and to have full authority over all traffic personnel and rate making.

The New York financiers accepted the traffic manager's recommendation, and instituted a highly centralized structure. The two key departments, traffic and transportation, remained completely separate and their managers reported directly to officials at the Chicago headquarters. The president and vice president personally co-ordinated the activities of the various departments.

The financiers preferred the centralized structure because it was less complex. By requiring fewer administrative personnel, it was less expensive than the system proposed by the acting general manager. Under centralization the traffic manager could quickly adjust rates and shift equipment to meet competition. But the centralized management blurred the distinction between routine operational matters and broad entrepreneurial decisions. In the long run top executives tended to get so involved in day-to-day problems that they had little time to devote to setting over-all policies and strategies. This defect, however, was not immediately apparent, and most of the railroad reorganizations supervised by J. P. Morgan in the 1890's followed the Illinois Central's example. These included

[57] Minutes of the "Board" formed to write up the Illinois Central's new "classification" or organization manual, held June 21, 1889 (Illinois Central Railroad Archives, Newberry Library, Chicago).

[58] T. J. Hudson to J. C. Welling, October 5, 1889 (Illinois Central Railroad Archives, Chicago).

the Erie, the Reading, Chesapeake and Ohio, and the Southern railroads.[59]

The events of the period from the Civil War to the beginning of the twentieth century profoundly shaped railway management. The competitive wars and the consequent formation of giant interregional railroad systems created mature managerial structures that varied only slightly from those used by present-day industrial giants. The administrative patterns erected drew heavily on the experience of the 1850's. In certain respects the structure that Thomson gave the Pennsylvania in the post-Civil War era was but a grander version of the one he fashioned for the same railroad in the 1850's. In the 1880's the New York Central formed a family of companies controlled through Vanderbilt's officers. Finally the Illinois Central carried the development of the centralized pattern to its logical conclusion.

Some Broader Implications of the New Administrative Forms

But structure alone was only part of the story. Equally as important were the changes that structure implied, many of which deeply affected American economic and social life. Within the ranks of management there developed a clear-cut separation between the top executives — the financiers who directed the competitive strategy and who made other strictly entrepreneurial decisions — and the supervisors who oversaw operating divisions, traffic departments, and the like. In both top and middle management professional experience and judgment became essential. There was little room for the part-time merchant promoter who, as late as the 1860's, played an important part in the management of many systems. This profoundly affected the railroad's stockholders and their representatives on the boards of directors, in short the owners who became increasingly dependent upon the advice of their hired managers. This led to the almost complete separation between "owners" and management which is so characteristic of the modern industrial corporation where complex problems require the services of the professionally trained and experienced expert.

In a similar way the new railroad administrative structure sharpened the gulf between labor and management; in fact, the two classes became separated by barriers almost as formidable as those

[59] Edward G. Campbell says little about administrative reorganization in his *The Reorganization of the American Railroad System, 1893–1900* (New York: Columbia University Press, 1938), Chapter 5; but Poor's "List of Officers of Operating Railroads," in the 1898 edition of his *Manual of the Railroads,* clearly indicates the centralized structures of these systems.

between the officer and the enlisted man in the military service. The expertise demanded of railroad financiers, rate makers, and other executives made it unlikely that a man without formal education who started out as a fireman or brakeman would rise to become a railroad president. The workmen themselves recognized this. Writing in 1889, Charles H. Salmons, a representative of the railroad brotherhoods, asserted that "the conditions of the trainman's life are hard. If he develops into a man of business or if he becomes a manager of great enterprises, it is in defiance of his surroundings." [60]

The painful, and at times hesitant, recognition that the traditional American Horatio Alger dream did not apply to railway labor led directly to the formation of the first really modern, national, skill or craft unions. The Brotherhood of Locomotive Engineers, oldest of the railway unions, began in 1863 when a group of Michigan Central enginemen organized to protect themselves from a "tyrannical" master mechanic.[61] By the 1870's several national brotherhoods had been formed, and were attempting to win better pay and working conditions for their men. In addition the unions provided their members with low-cost accident insurance, a necessary step because of the highly dangerous working conditions and the inadequate compensation for injured employees provided by the corporations.

Railroad administrators strongly opposed the unions. The Burlington's John N. A. Griswold saw the strike against his road in 1888 as "not a question of money, but as to who shall manage the road." [62] But despite opposition the brotherhoods grew, and by 1900 they were accepted by most railroad executives. By that time, too, Congress had passed the Erdman Act, which provided governmental machinery for mediation and, if necessary, compulsory arbitration in railroad-labor conflicts.[63] Thus, on the railroads, where modern management and modern labor unions first appeared, there

[60] C. H. Salmons, *The Burlington Strike: Its Motives and Methods Including the Causes of the Strike, Remote and Direct, and the Relations to it of the Organizations of Locomotive Engineers, Locomotive Firemen Switchmen's M.A.A., etc.* (Aurora: Burnell and Ward, 1889), p. 14.
[61] Donald L. McMurry, *The Great Burlington Strike of 1888: A Case History in Labor Relations* (Cambridge: Harvard University Press, 1956), p. 29.
[62] Quoted in Cochran, *op. cit.*, p. 180.
[63] The Erdman Act, passed in 1890, provided for mediation by the Chairman of the Interstate Commerce Commission and the Commissioner of the Bureau of Labor. The Act was little used, however, until Theodore Roosevelt's second term. From 1906 through 1916 sixty-one railroad labor controversies were settled under the terms of the Act.

also developed methods of bringing these two powerful groups together which were not used by the rest of American business for more than a generation.

Some Generalizations about Innovative Industries and Their Impact on Administration

There can be little doubt that the impact of the railroads on American administrative practice has been great. The question inevitably arises as to whether future innovative ventures will bring similar changes or at least ones equally as dynamic. Is it possible, for example, to draw an analogy between two such diverse, but admittedly innovative efforts as railroads and space? We flatly reject the idea that historical analogies can be constructed which will predict future trends. We do believe, however, that some valid generalizations can be abstracted from the railroad experience which can be made the basis for hypotheses, and which may aid in the analysis of other innovative industries.

It appears, using the railroad experiences as a guide, that the following hypotheses can be asserted. First, that innovative industries create a need for new administrative patterns, but that these requirements do not become obvious during the first few years of the industry's existence. As a corollary, therefore, one may add that the original promoters use traditional methods to manage their new ventures. Certainly, early railroad entrepreneurs did not regard their businesses as unique. Any analysis of the space effort might start by examining the origins of its present administrative system. Was it designed to conform with current governmental and industrial standards, and do the administrators regard their enterprise as presenting unique managerial problems?

A second axiom might read that operational crises within a new industry emphasize the need for, and also shape, initial administrative reforms. It has long been a legend that the prevalence of United States Military Academy graduates in the ranks of early railroad builders and managers was especially significant. The evidence, however, clearly refutes the idea that army administrative procedures influenced railroad practice. On the Western Railroad in Massachusetts, in the 1840's, neither West Point graduate and Chief Engineer George Whistler nor George Bliss, the Springfield lawyer-president, made any attempt to reform a traditional administrative structure until the railroad suffered serious operational breakdowns. And the advanced patterns which emerged in the 1850's on the Erie, the Baltimore & Ohio, and the Pennsylvania were in direct response to the challenges presented by a railroad five hundred miles long. There can be little doubt that the Western Railroad's troubles, which received wide publicity in trade magazines such as the *American Railroad Journal*, provided the real basis for constructive thought about the management of long railroads. What kinds of operational crises, if any, will mar the space effort are still a matter of conjecture. It might be worthwhile, however, to examine failures of past space probes to determine if novel problems have occurred and, if so, whether these have exercised an effect on administrative practice.

A third hypothesis is that in the long run administrative patterns, although deeply influenced by operational and technological considerations, are also profoundly shaped by the outside environment. After 1865, forces within the American economy threw the railroads into cutthroat competition, which in turn emphasized the non-operational problems of rate making, finance, and expansion through construction and purchase. Thus the operational managers were thrust aside and control gravitated to financiers, who were chiefly responsible for the administrative structures that evolved between 1865 and 1900. The financiers by and large neglected administration, but ultimately most railroads reorganized their management in accord with the centralized departmental concept of the Illinois Central or the decentralized divisional structure of the Pennsylvania. The experience of the late nineteenth century suggests two corollaries to the third axiom. The most obvious is that no single type of structure is inevitable. And second, that men trained in the daily operation of the unit tend to have superior administrative skills to those who have no such experience. The care and thought with which Thomson managed the Pennsylvania contrast sharply with the haphazard way in which the Illinois Central achieved its structure. The foregoing observations seem especially relevant to the space program since it is certain that outside pressure, from politics and fiscal limitations, for example, will constantly affect that program. In the years ahead space may draw its leadership from such varied fields as politics, the scientific community, and the military. Any analysis of the administrative development would do well to compare and contrast the quality of leadership arising from these various sources.

Finally, it seems evident that new administrative structures alter fundamental relationships between labor and management. The nineteenth-century railroad administrative practices created a sharp cleavage between the managerial elite and the ordinary workers.

This resulted in the rise of the modern craft union and of collective bargaining. The space effort, though potentially a vast source of employment, will involve very little unskilled or semiskilled labor. This raises some interesting questions. Will a highly professional working force cause administrative reforms, and if so, what will be the role of the labor union?

The end of the nineteenth century saw the railroad network stabilize. The industry ceased to be innovative, and its management became routine. Many men who received their initial experience with the railroads went on to administer other large corporations. And bankers such as Morgan applied concepts that they learned in railroad reorganizations to other industrial ventures such as United States Steel. But the basic administrative problems which the railroads solved are still central to most large enterprises. Space, therefore, starts with advanced managerial concepts many of which have been tested and refined by large organizations since the 1850's. What alterations to present administrative practice the complex problem of space exploration will produce remain for future historians to study. It is our hope, however, that the preceding study of nineteenth-century railway managerial practices will provide helpful insights into the problem of analyzing trends in the administration of giant enterprises.

6

The Social Impact of the Railroad

Thomas C. Cochran

In this chapter the term social will be narrowed arbitrarily to exclude economic and business effects of the railroad, because other chapters cover these areas, but the scope will be extended sufficiently to cover certain psychological effects. This use of "social" conforms to that usually employed by historians in defining the field of "social history." For purposes of analysis it is convenient to divide the social effects of the railroad into three categories: demographic, institutional, and social-psychological.

Demographic Effects

The major demographic effects of the railroad were increased spatial mobility and redistribution of population.

Spatial mobility, or migration, has been one of the major influences in shaping American culture. As Everett Lee has demonstrated, nearly all of the characteristics attributed by Frederick Jackson Turner to the influence of the frontier are also typical of the behavior required for successfully playing the role of in-migrant, whether to rural or urban communities.[1] Hence, any factor affecting the nature of migration is an important influence on social change.

Each improvement in inland transportation presumably increased the average range of migration. Even in the late colonial period newcomers who were not bound to labor usually found land within a hundred miles of the port of debarkation; and freed servants, farmers' sons, or previous settlers selling and moving, probably went less than a hundred miles. Moves from the Lancaster, Pennsylvania,

[1] Evertt S. Lee, "The Turner Thesis Reexamined," *American Quarterly*, XIII (Spring, 1961), 77–83.

area to a "western" farm across the Susquehanna, or from eastern to western Massachusetts, may tentatively be assumed to be modal distances. There were, of course, longer journeys by certain groups, such as the Scotch-Irish immigrants of the 1740's, who in some instances moved from western Massachusetts to western Pennsylvania, but it seems safe to regard these as exceptional.

The river steamboat, and the opening of western lands previously controlled by Indians, extended the range of migration along the relatively small areas near waterways. Migrants who could afford to cross the Appalachians, an exhausting and expensive journey, might go considerable distances down the Ohio River. The same was true of those reaching Buffalo and sailing westward on the Great Lakes. But again, this long-distance migration was probably a small part of the total. The hard-surfaced road, after the mid-1790's, facilitated local movement in the populous East, and on a few roads penetrating into the trans-Appalachian West. But the median, as distinct from the average distance of movement may not have been much greater than in the colonial period.

For all periods, it is impossible from published records to ascertain exact figures on the average length of travel of the population moving into a given area, such as a county, before and after the coming of the railroad. Place of birth is recorded in the censuses, but not place of last residence. Study of the "birthplace by county" data for Iowa in the censuses of 1870 and 1880 has revealed four propositions about the effect of railroad construction on population distribution:[2]

1. The railroads completely overshadowed other routes of transportation as determinants of migration.

2. They stimulated greatly the growth of areas connected with centers of population; this stimulus was most pronounced in the case of the foreign born, intermediate in the case of persons born in other states, and least in the case of persons born in the state.

3. The growth of the foreign-born population was due, in the

[2] Based on research by G. Putnam Barber, University of Pennsylvania. The four propositions can be clearly shown in the data for Iowa by use of what Barber calls the index of migration impact. This index is constructed by taking the ratio of percentage increase in the population from a certain birthplace in a given county to the increase for the state as a whole. The index fluctuates around 100; values greater than 100 indicate that the county gained population from a given birthplace faster than the state as a whole. Strict interpretation of the index would require that one assume that the force of mortality, and the propensity to remigrate, were equal in their impact on persons of all birthplaces in all counties. This strict assumption can be relaxed, however, if all that is necessary is a general impression of the effects of differing patterns of migration.

THE SOCIAL IMPACT 165

case of the Germans, to new migrants to the area, and in the case of the Irish, to "drop-offs" from the construction crew. (Other nationalities were not sufficiently represented in the population of Iowa at the time to permit such analysis.)

4. In the first few years following the construction of a railroad, the population of a county received a relatively larger share of migrants from more distant states than either before or after that time.

As a result of Barber's research and Curti's study of Trempealeau County, Wisconsin, it appears that the extension of both the median and average length of each migration as a result of the railroad cannot be documented.[3] The fact that the secular trend of foreign immigration was rising at the same time (1845–1890) that the railroad was penetrating the interior strongly supports the deduction of an over-all increase in distance of migration but does not touch on the problem of the length of each specific move by the previously resident population.

Another effect of the railroad operated to increase the usual distance of migration. From 1850, the major direction of migration was not from south to north or east to west but from country to city. The migrant leaving a farm to go to a city must, in general, have selected his destination on the basis of imagined opportunity.[4] This means that he would not necessarily have gone to the nearest city, for instance from a farm in central Vermont to Rutland, but rather to a city that was expanding rapidly and offered many job opportunities, such as Buffalo, some three hundred miles away.

A minor result of the coming of the railroad was the creation in the trans-Mississippi West of large pockets of immigrants from the same country where the native language and customs would be preserved for at least a generation. Some groups, such as the Pennsylvania Germans of the colonial period, and the Cincinnati, St. Louis, and Milwaukee German concentrations established by water transport, had preceded the railroad. But the latter spread such pockets into the rural areas of the Mississippi and Missouri valleys.

This conclusion leads directly to another major aspect of the railroad impact: the location of population in areas that had not been agriculturally profitable with previous forms of transportation. The railroad with its ability to spread in most directions, particu-

[3] Merle E. Curti and Others, *The Making of an American Community* (Stanford: Stanford University Press, 1959), pp. 62–64.
[4] See Carter Goodrich and Others, *Migration and Economic Opportunity* (New York: Columbia University Press, 1935).

larly over the level plains, could open up more of the country than could water transportation.

It was, of course, greatly inferior in this respect to motorized transport over hard-surfaced roads. But even without the competition of the railroad, motor vehicles would probably not have been economically successful before the late 1860's or 1870's. Such transportation required light steam or gasoline engines, and while experiments with steam vehicles were under way in England in the early 1860's, and the development of the internal combustion engine in Germany from 1865 to 1885 would no doubt have been greatly stimulated, commercial adaptation would have taken time. Macadam roads, on the other hand, were technically adequate and could have been built rapidly in most parts of the United States.[5]

The demographic effect of the railroad was, therefore, a more uniform deployment of population over land resources than would have been possible previously, but a distribution that still did not give farmers the best conceivable combination of the factors of fertility and nearness to the consumer. Long fingers of population stretched along the railroads and left land more than about a dozen miles from the track unused, or sparsely inhabited by semi-self-sufficient, noncommercial farmers or hillbillies — even though the hills might be very low.[6] By 1890, in the area from the Ohio Valley north and west through the first tier of states beyond the Mississippi River, main and branch railroad lines made such a fine network that in fertile areas nearly all of the land was within twelve miles of a station; but in the mountains of the South and the plains of the trans-Mississippi West this was not the case.[7]

Population was literally drawn by the railroads into the "middle border" through construction work, sale of land on credit, eastern and foreign advertising, and offers of free or cheap transportation.[8] Certainly, up to 1890, the population of the states between the Mississippi River and the Rockies increased much faster than would have been possible without the railroad. Whether all of this increase was economically or sociologically desirable is another question.

[5] See further discussion of this problem in Robert W. Fogel, *Railroads and American Economic Growth: Essays in Econometric History* (Baltimore: The Johns Hopkins Press, 1964).

[6] See Abram Kardiner, *Psychological Frontiers of Society* (New York: Columbia University Press, 1945), pp. 294–295.

[7] *Grant's Bankers and Brokers Rail Road Atlas, 1890* (New York: A. A. Grant, 1887).

[8] See James B. Hedges, *Henry Villard and the Railroads of the Northwest* (Cambridge: Harvard University Press, 1930); and Richard C. Overton, *Burlington West* (Cambridge: Harvard University Press, 1941).

Instead of filling in the more easterly states, farmers were led to cultivate the staples on large acreages at immense distances from their markets, to live in difficult and uncertain climates, and to endure relative social isolation over most of the year.[9]

By making it easier to reach interior cities, the railroad undoubtedly increased the spatial mobility of labor. This was particularly true in many small to medium-sized cities that were not on navigable waterways. The effect on the major industrial concentrations, which were served by water, would be hard to estimate. Furthermore, many factors may have been as important as transportation in determining these internal migrations. The desire of the southern Negro to reach freedom before the Civil War, and thereafter to reach greater opportunity, operated in spite of nonexistent or poor public transportation. Immigrant workers were drawn to cities where they had fellow countrymen, skilled workers to where there was a special demand for their abilities, and partially employed workers to where there was full-time work, all without much apparent regard to transportation difficulties.[10] On the other hand, workers with families, who owned homes or were well located in a city, could not readily be drawn as far as Chicago from Milwaukee by wage differentials as high as 25 per cent.

The railroads sought to stimulate immigration in order to supply labor for their own construction needs, and to provide farmers and other businessmen for the development of the areas the companies served. Importation of unskilled labor under contract to work for repayment of transportation costs involved too much risk.[11] The United States was too big and jobs were too plentiful. Instead, the railroads advertised for workers in Europe and in the eastern ports, and provided cheap transportation from the coast. The Castle Garden Labor Bureau, supervised by the State of New York, was probably the chief mid-century source of immigrant labor for the railroads.[12] The *Railroad Gazette* noted in 1883, the peak year of immigration up to that time and also a peak year in railroad building, that "unless the rate of immigration continued high construction would be handicapped." [13]

[9] See Walter A. Terpenning, *Village and Open Country Neighborhoods* (New York: The Century Co., 1931), pp. 53–54.
[10] This subject requires much more research. See Charlotte Erickson, *American Industry and the European Immigrant, 1860–1885* (Cambridge: Harvard University Press, 1957).
[11] *Ibid.*, p. 73.
[12] *Ibid.*, pp. 95–96, 118.
[13] Quoted in *ibid.*, p. 76.

The shift of population in both rural and urban regions that resulted from the railroad also relocated industry and redirected trade. Many important effects of these changes were obviously economic, but from the demographic standpoint the process was often a snowballing one. The more the railroad concentrated population, the more people seeking opportunity came to these points of concentration. Thus the process of depopulating rural areas which goes far back in American history, was stimulated by the railroad.

In the nineteenth century railroad rates were not uniform, but were based on what the traffic would bear. Between points where there was competition between railroad companies, or with water carriers, rates were kept low, often below the level necessary to cover overhead costs and moderate dividends. To compensate for this the roads charged higher than average rates between points where there was no competition. As a result, manufacturing and wholesale trade were concentrated at the competitive or junction points, and the more carrier competition there was the more likely that rates would be low. If only two carriers were competing, they might form a "pool" that set rates and apportioned traffic, but between three or more carriers pools were much more difficult to maintain.[14]

Another factor tending to make big cities still bigger was that the short, early railroads frequently ended in such cities, and might fail to connect directly with other roads or to use track of similar gauge. The resulting transshipment made the points economically advantageous for processing plants.

Thus, businesses and their workers were drawn from small city locations to the major railroad junction points. In spite of lower fertility rates the major cities grew much faster than the rest of the population. By 1900 the pattern of what would now be called metropolitan areas was, in its new parts, a creation of the railroad. Chicago, the greatest of the cities built by the railroad, grew from 30,000 people in 1850 to 1,100,000 in 1890.

Among the minor effects of the increased spatial mobility brought about by railroads was the alteration of Indian relations. The railroad increased the sense of grievance of the plains tribes since it encouraged organized hunting parties which virtually exterminated the buffaloes, and brought white men into areas awarded by treaty

[14] See Julius Grodinsky, *The Iowa Pool: A Study in Railroad Competition, 1870–1884* (Chicago: University of Chicago Press, 1950); and Thomas C. Cochran, *Railroad Leaders* (Cambridge: Harvard University Press, 1953), pp. 170–172.

to the Indians. But it also ended the possibility of effective Indian military resistance. The United States government abandoned the fiction of dealing with the Indian tribes as sovereign nations by treaty and moved to the policy of creating restricted reservations, of granting federal aid, and eventually of granting citizenship.

Institutional Effects

Family

Initial migration from Europe, and continued migration within America, prevented the extended family, or large interdependent kinship groups, from developing in most parts of the country. Exceptions were most notable in parts of the southern seaboard and in poor rural areas. Here the more enterprising members of the family left to seek better opportunities, and those who remained sought security through emphasizing kinship. This pattern was present to a minor extent in all areas. Cities apparently had some solid residues of kinship groups that stayed in place from generation to generation while in-and-out migrants washed over them in waves.[15] But in the cities, the security afforded by the extended family was much less than in the country, and the bonds between its members correspondingly weaker. These parents, grandparents, brothers, sisters, aunts, uncles, and cousins living apart, but nearby, might be called a fringe of the nuclear family, rather than an extended family of the type prevalent in the non-European world.

While, as argued in the previous section, the railroad increased the median distance of each single move, it also made it possible to spread the nuclear fringe more widely without breaking the fabric. Before the railroad, a son who moved his nuclear family fifty miles away from the paternal homestead was living in another area. Visits taking two days in each direction would be infrequent. With rail service the trip could be made in two hours, and visiting and co-operation between relatives could be preserved. Even within a neighborhood area six or eight miles in diameter, two or more railroad stations might serve to knit a kinship group more closely together.

Education

The railroad was probably a major factor in developing higher education in engineering. At the time that Rensselaer Polytechnic

[15] Sidney Goldstein, *Patterns of Mobility, 1910–1950: The Norristown Study* (Philadelphia: University of Pennsylvania Press, 1958), pp. 214–215.

Institute was founded in 1824, the Military Academy at West Point was the only college offering advanced technical studies. The founder of the Institute, Stephen van Rensselaer, thought that training in chemistry for use in agriculture would be the school's chief function, but the demand for engineers to build bridges and railroads soon moved the curriculum in that direction.

Just before the great railroad construction boom from 1849 to 1854, Amos Lawrence gave Harvard $100,000 for a scientific school. Like van Rensselaer, Lawrence was not specifically interested in developing railroad engineers; but Lawrence Scientific School, opened in 1847, inevitably moved toward an emphasis on the civil engineering required for railroad construction. Meanwhile, Yale had established a scientific department which became Sheffield Science School in 1854. By the time the railroad boom leveled off, Dartmouth, Michigan, and other universities had opened engineering schools or departments, and Brooklyn Polytechnic Institute (1854), Cooper Union (1859), and Massachusetts Institute of Technology (1860) had been founded as independent technical schools.

As noted in the case of Rensselaer and Lawrence School, growth of railroad demand was not the sole factor in the spread of engineering education, any more than missiles and space developments were the sole cause of the great mid-twentieth-century upswing in scientific education; but without the pressure of the roads during their major building periods, from 1849 to 1857 and 1867 to 1890, the growth of such education would surely have been slower.

Business

While a few canals like the Delaware and Hudson, and the Lehigh were operated by large corporations, the railroad was the early developer of the institutions of big business. These included one or more levels of middle, or bureaucratic management; and operating policy was formulated by professional executives rather than by the principal stockholders. Inside, the railroad companies' new management forms, described by Chandler and Salsbury, developed rapidly; outside, the companies' railroad administrators added substantially to the growing upper-income group.

Inevitably, the long-range planning necessary to a railroad, and the many contacts with the public and with government, gave railroad managers an unusual awareness of their position in society. William H. Vanderbilt's, "The public be damned," was not a declaration of independence but rather the complaint of a weary man tired of giving news releases on problems of public and governmental relations. More representative is his statement: "We must be conservative and keep the public with us." [16] It took a century to develop a reasonably full conception of what the status of professional manager in a big company involved, but the process of elaborating this conception was well advanced among railroadmen by the 1880's.

Undoubtedly, this new social group put an increased emphasis on efficiency, both technologically and socially. In trade and manufacturing, wealth came from market profits, often of a chance variety; while for the professional railroad manager wise expansion and efficient operation, rather than profit, per se, were the economic aims. As President J. C. Clarke of the Illinois Central wrote in 1885: "it is better to show moderate poverty than great success," since the latter would provoke more competition and government regulation.[17] Determining the effects on society of largely increasing the number of prominent citizens with such bureaucratic or professional views will require further research. In some instances the effects might be analogous to introducing five thousand space scientists into a profit-oriented community such as Houston, Texas.

The railroad popularized investment in stocks and bonds other than those of banks or governments. Every important railroad was "publicly" financed. By 1855, the securities of roads like the New York Central or the Pennsylvania were held by thousands of investors. Higher interest rates on bonds than those paid by sound governments, and the chance for appreciation in railroad stocks, led more prosperous business or professional men to buy railroads on the security exchanges. Superficially the rise of security capitalism may seem to have been an economic effect, but it probably had important social implications. The middle-class security holders formed a generally conservative and powerful social group resistant to radical experiments or change. No country in which this group is strong has moved from capitalism to socialism.

Government

Americans of the early nineteenth century did not fear their governments. The states, particularly, were seen as useful mechanisms for advancing the general welfare, and like other people in an early stage of economic development, the citizens of some states

[16] Cochran, *op. cit.*, p. 157.
[17] *Ibid.*, pp. 127–128.

believed that for many of the needed facilities such as banks, canals, railroads, or universities, public enterprise was more practical than private enterprise. For example, a commission on canals in New York, headed by the wealthy and conservative Gouverneur Morris and composed largely of wealthy men, reported in 1811 that "by the tests of efficiency, economy and financial capacity, private enterprise was incapable of building the Erie and Champagne canals." [18]

When Massachusetts was faced with the competitive need for building a railroad, it was thought that private capital would not venture into so hazardous an enterprise "without the pledge of more exclusive privileges than it would be expedient for the legislature to grant." "The important interests of the people," Governor Lincoln held, "can only be preserved, and the honour and prosperity of the State promoted, by a system of Governmental enterprise." [19] Prompt state action was delayed in Massachusetts by fear of the cost. Some said it would be "little less than the market value of the whole territory of Massachusetts." [20] Ultimately the state paid over half the cost of the Western Railroad.

The pioneer railroads were generally either government owned or deeply involved in financing by governmental units. They also required franchises and the ability to condemn property. Thus, like canals, they were closely tied to state politics. On some roads, such as the New York Central, railroad executives were also legislators or high state officials; but, in any case, a railroad company needed professional lobbyists. The lobbyist was usually a politically active lawyer with a practice at the state capital. Robert S. Hunt writes of the Wisconsin antilobbying law of 1858 that it "did not prevent the roads from dominating the Wisconsin legislature throughout most of the period from 1850 to 1890." [21]

Some roads like the Pennsylvania and the Michigan Central remained under state ownership until the late 1840's. Two roads, the Western and Atlantic of the State of Georgia and the Cincinnati and Southern, although leased to private companies, were owned by the state governments until the late nineteenth and early twentieth centuries, respectively. However, there was little government construction after 1857. Of the pioneer period to 1890, Hunt says: "The

[18] Nathan Miller, *The Enterprise of a Free People* (Ithaca: Cornell University Press, 1962), p. 32.
[19] Oscar and Mary F. Handlin, *Commonwealth* (New York: New York University Press, 1947), p. 186.
[20] *Ibid.*, p. 187. There is more on this theme in text and notes.
[21] Robert S. Hunt, *Law and Locomotives* (Madison: State Historical Society of Wisconsin, 1958), p. 171.

notion that railroad building was a public function, widely accepted at the beginning was faint indeed at the end." [22]

In railroading, therefore, business and government were in partnership during the early period, much as in the mid-twentieth-century atomic power industry and certain defense industries. Usually state and municipal governments were bondholders, but occasionally, as in the Camden and Amboy Railroad of New Jersey and the Western of Massachusetts, the state was a common stockholder. Three of the nine directors of the Western were appointed by the state.[23] In return for a gift of a thousand shares of common stock, New Jersey gave the Camden and Amboy a monopoly on passenger traffic across the state; if a competitor was chartered, the state had to surrender the thousand shares to the company.[24]

Many other types of aid by the states and municipalities in both the East and the West could be described.[25] But the point to be made here is that the railroad as an innovating agency depended heavily on government encouragement. Only a few short lines between populous eastern cities were wholly financed by private capital.

The federal government became important to the railroads first for mail contracts, and after 1850 for land grants. From 1850 to 1862 the federal grants were made indirectly through the states, but from 1862 to 1870 large grants were made by the federal government directly to private corporations for potential transcontinental roads. The Central and Union Pacific companies were also aided by some $50,000,000 subscribed by the United States for second mortgage bonds, and two other roads received some minor aid of this type. In the land grants a provision was always included for reduced fares for government freight and the transportation of troops.

These policies made strong representation in Washington desirable for all mail-carrying and land-grant roads. In some cases, such as that of the Union Pacific, lobbyists dispensed railroad stock in order to gain friends in Congress. Meeting the conditions for transfer of land from the government to the railroad, locating lands, and

[22] *Ibid.*, p. 167.
[23] Cochran, *op. cit.*, p. 18.
[24] John W. Cadman, Jr., *The Corporation in New Jersey* (Cambridge: Harvard University Press, 1949), pp. 54–55. No competitor was chartered before the Civil War.
[25] See F. A. Cleveland and F. W. Powell, *Railroad Promotion and Capitalization in the United States* (New York: Longmans Green, 1909); and Harry H. Pierce, *Railroads of New York: A Study of Government Aid, 1826–1875* (Cambridge: Harvard University Press, 1953).

getting rights-of-way through Indian territories kept the officers of the growing transcontinental roads in close touch with their men in Washington up into the 1880's. Chester A. Arthur, later President of the United States, was a lobbyist for Northern Pacific in the 1870's.[26]

American courts, always guided to a degree by precedent, tend to be conservative. For many decades it is possible to evade major issues. Wisconsin law, writes Hunt, "failed to recognize the need for anything more than the time-tested techniques of an agricultural society."[27] In Massachusetts, "looseness of control, liberal incorporation and freedom of lines to use each other's tracks encouraged the tendency to allow all to proceed at their own risk without favoritism, that is, without direction from the government."[28]

Yet little by little corporations, and railroads in particular, by forcing the courts to give "specific interpretations in case after case gradually undermined and transformed the law."[29] In general, change was in the direction of more sovereignty for the corporation and less control under the assumptions of the common law.[30]

While the railroad's effect on common law and judicial process was both protean and subtle, and has had little study, its effect on statute law was obvious and important. Railroads were in fact local monopolies whether or not they were granted such privilege in their charters; and to lay out their routes they required the right of eminent domain.[31] It was inevitable that private corporations possessed of such powers would be regulated by the states that created them. New England had the largest amount of track in relation to its area, and first had to meet most of the problems of railroads as public utilities. Fortunately the history of New England railroads has been studied in detail by Edward C. Kirkland.

Eminent domain had been used by transportation companies and public utilities, such as waterworks, before the coming of the railroad. But the rapid growth of railroad companies, and public excitement about the powers of monopolistic corporations in the Jackson period brought the right into question in the Middle Atlantic and New England states. In 1833, a Massachusetts law prescribed the method by which railroads might exercise eminent domain, and in 1844, New Hampshire provided for appeal to a board of three

---

[26] *Billings Correspondence* (Northern Pacific Archives, St. Paul, Minnesota).
[27] Hunt, *op. cit.*, p. 171.
[28] Handlin and Handlin, *op. cit.*, p. 193.
[29] *Ibid.*, p. 149.
[30] Hunt, *op. cit.*, p. 171.
[31] The first three Massachusetts roads were granted thirty-year monopolies.

commissioners who could fix valuations and assess damages.³² But the more general attitude, which prevailed in the three Middle States, was that charters were self-enforcing subject only to appeal to the Courts.³³

While for any given area the exercise of eminent domain was a passing problem, that of rates and conditions of service remained. The first state railroad commission resulted from the question of whether a railroad, the Boston and Providence, was open to use by separately owned connecting lines. The practical issue was settled when the Boston and Providence bought the line in question, but the issue led, in 1839, to the creation of a Rhode Island Railroad Commission of five members.³⁴

Gradually the New England states and New York found it impossible to handle railroad problems by charters and legislation alone. "By the fifties," writes Professor Kirkland, "it was apparent that governors, attorney-generals, legislatures and their committees, judges, and county commissioners were not able to deal continuously, expertly and successfully with a powerful, expanding, novel, railroad interest." ³⁵ Connecticut established a railroad commission in 1853, New York in 1855, Maine in 1858, and Vermont a single commissioner in 1855.³⁶ These early commissions were more concerned with condemnation, connections, and public safety than with rates.

By the late 1860's spread of the railroads into western areas involving long hauls and transshipment, combined with declining commodity prices and less rapidly falling railroad charges, tended to focus public discontent on rates. The first of the post-Civil War commissions, however, was an outgrowth of a number of considerations in Massachusetts, rather than of western freight-rate problems. This commission, created in 1869, had broad responsibilities for supervision, including methods of accounting and rates, but no

---

³² Edward C. Kirkland, *Men, Cities and Transportation: A Study in New England History* (2 vols., Cambridge: Harvard University Press, 1948), Vol. I, pp. 274–275.

³³ See Lee Benson, *Merchants, Farmers and Railroads* (Cambridge: Harvard University Press, 1925), p. 2; Cadman, *op. cit.*, pp. 222–224; Louis Hartz, *Economic Policy and Democratic Thought* (Cambridge: Harvard University Press, 1948), pp. 69ff.

³⁴ Kirkland, *op. cit.*, Vol. II, p. 234.

³⁵ *Ibid.*, p. 232.

³⁶ *Ibid.*, pp. 233–236. For New York, see Benson, *op. cit.*, pp. 6ff. The New York Commission was ended by a corrupt deal in 1857, and not recreated until 1882.

power except to report and recommend.³⁷ In Illinois, a center of violent antagonism toward the railroads, the Chicago Board of Trade, aided by farm organizations, secured a law in 1871 that fixed maximum rates and prohibited discrimination but set up a commission that lacked power.³⁸ Before 1880, the other states of the northern Mississippi Valley and California enacted similar laws. Only Missouri gave its commission the power to adjust rates, rather than merely to enforce maximums set by law.³⁹

Taking a broad view of the twenty-year struggle between midwestern railroads and public authorities, it can be said that each local group of shippers needed the railroads more than the latter needed those particular shippers, and each year the railroads gained more subtle political leverage.⁴⁰ While the United States Supreme Court upheld state regulation in *Munn* vs. *Illinois*, in 1876, the railroads had gained friendly majorities in some of the legislatures by this time and were exerting effective influence on the commissioners. As President Henry B. Ledyard of the Michigan Central wrote in 1883, "Where there are commissioners to stand between the railroads and the public much dissatisfaction can be avoided and many things made plain." ⁴¹ Another factor working in the railroads' favor was increasing technological efficiency which made it possible to earn profits from lower and lower rates.

While the roads were adjusting to state legislation, and vice versa, there was a movement in Congress for federal action. In 1874, the Windom Report, by a congressional commission, urged federal construction of a rail and water system to provide what would later be called "yardstick" competition for the private roads.⁴² From 1878 to 1885, Representative J. F. Reagan of Texas annually introduced regulatory bills. Charles E. Perkins of the Burlington Railroad was by then convinced that "the public *will* regulate us to some extent — we must make up our minds to it," and other presidents struck the same note.⁴³ By 1886, a number of railroad executives welcomed regulation by a federal commission as a means of preventing cut-throat competition.⁴⁴

---

³⁷ Kirkland, *op. cit.*, Vol. II, pp. 238–239.
³⁸ Benson, *op. cit.*, p. 25.
³⁹ Sidney F. Miller, *Inland Transportation* (New York: McGraw-Hill, 1933), pp. 79ff; Robert E. Riegel, *Story of the Western Railroads* (New York: The Macmillan Co., 1926), pp. 143ff.
⁴⁰ Cochran, *op. cit.*, pp. 189–196.
⁴¹ *Ibid.*, p. 191.
⁴² Riegel, *op. cit.*, p. 142.
⁴³ Cochran, *op. cit.*, p. 197.
⁴⁴ Gabriel Kolko, *Railroads and Regulation, 1877–1916* (Princeton: Princeton University Press, 1965), pp. 26ff.

In the course of its first half century of relations with state and federal governments the railroad had introduced most of the present-day practices of lobbying and public relations. Passes were issued to ministers, editors, and politicians; influential legislators, senators, or public executive officers were employed while in or out of office to plead the railroad's cause; editors and free-lance journalists were hired to write articles; and, on occasion, cash payments or securities were distributed to key figures.[45] While it is obviously impossible to assess the total effect of such activities, it is significant that the United States became the only major industrial nation without state-owned railroads.

Social-Psychological Effects

Without special research by psychologists in historical records it is impossible to do more than note certain obvious psychological effects of the railroad.

The railroad raised the levels of expectation in relation to material success in many communities, and altered both geographic and cultural horizons. In its daily life the prerailroad rural community represented a high degree of self-sufficiency. Only in the neighborhood of big cities was community interest sufficient in New England to support the construction and maintenance of good hard-surfaced roads.[46] Lewis E. Atherton sees the prerailroad midwestern town as characterized by "a low standard of living, relative lack of specialization, and freedom from outside control." [47] When Emerson wrote of Massachusetts, "From 1790 to 1820, there was not a book, a speech, a conversation or a thought in the State," he should have added, according to Professor Kirkland, "that there was not a railroad. For the railroad, even though it may not have opened wider prospects, at least revealed different ones." [48]

The initial effects of the railroad, often merely of the planning for a railroad, have been extensively discussed. Describing a meeting to promote a railroad in western Pennsylvania, the editor of the *Pittsburgh Daily Gazette and Advertiser* wrote: "We believe no other subject matter whatever, unless it was akin to a revolution when men felt their religions or civil liberties were at stake could

[45] Cochran, *op. cit.*, pp. 184–189, 193–196.
[46] Kirkland, *op. cit.*, Vol. I, p. 37.
[47] Lewis E. Atherton, *Main Street on the Middle Border* (Bloomington: University of Indiana Press, 1954), p. 42.
[48] E. W. Emerson and W. E. Forbes, eds., *Journals of Ralph Waldo Emerson* (Boston: Houghton Mifflin, 1909–1914), Vol. VIII, p. 339. Kirkland, *op. cit.*, Vol. I, pp. 92–93.

have collected such a gathering."[49] The people of Oswego, writes Professor Pierce, were "unanimous for anything in the form of a railroad, whether it goes crooked or straight they seem to have no care."[50] In Massachusetts the influence of the church was invoked by the promoters of the Western Railroad. Ministers were asked to discourse "on the moral effects of Rail-roads."[51] One of the most interesting accounts of the feverish enthusiasm, overoptimism, and broader perspectives generated by the coming of the railroads is in R. Richard Wohl's, "Henry Noble Day."[52] The coming of a branch-line railroad to Hudson, Ohio, stimulated visions of [53]

> a whole sub-system of other roads all centered on Hudson. . . . For Henry Day himself the dream had become a sacred reality. He was able to anticipate its completion while the first few spadefuls of earth were taken out for grading. He proceeded therefore to enact the logic of his expectations. He began to create a network of businesses in Hudson to service the demands arising out of the railroad building boom and to cash in on the enlarged market which would result once the roads were completed.

The resulting boom and bust was painful, but undoubtedly educational as well.

Railroads brought city newspapers to rural towns with advertisements of bargains in clothes and home fittings, and of amusements not thought of before by country people. Now on a Saturday the farm family could take a one-day cheap excursion train to the biggest nearby city and live for four or five hours in a new world.[54]

New contacts for rural people and a broader sphere of influence for urban institutions undoubtedly stimulated innovation. Bringing the farmer to the city could of itself be important. In the United States, writes H. G. Barnett, "the cultural differences between rural and urban life patterns that appear at farmers' markets are not insignificant from the standpoint of change."[55] The disruption of normal rhythms of activity by local booms, such as that of Hudson,

---

[49] Louis Hartz, *op. cit.*, pp. 11–12. From paper of March 18, 1846, cited in *Western Pennsylvania History*, Vol. XIII (1930), p. 252.

[50] Pierce, *op. cit.*, p. 42.

[51] Handlin and Handlin, *op. cit.*, p. 227.

[52] In William Miller, ed., *Men in Business* (Cambridge: Harvard University Press, 1952), pp. 153–192 (Torchbook edition, 1962).

[53] *Ibid.*, pp. 176–179.

[54] See Atherton, *op. cit.*, p. 230.

[55] H. G. Barnett, *Innovation: The Basis of Cultural Change* (New York: McGraw-Hill, 1953), p. 46.

Ohio, provided the individual with "auspicious circumstances for the emergence of new ideas, many of them predatory. . . . The microcosm in which he has lived has been destructuralized and is not habitable in that condition. He strains to give it some organization and some meaning, and in so doing he innovates or accepts definitions of the situation offered by others." [56]

To these normal incentives to innovation and change new American communities presented the added factor that, "Migrants and dispossessed populations are characteristically receptive to new ideas, whether these ideas are developed by their own members or suggested by outsiders." [57] The ideological dangers to the *status quo* inherent in the railroad were recognized by the school board of Lancaster, Ohio, which forbade the use of the schoolhouse for a meeting to promote a railroad on the basis that "such things as railroads and telegraphs are impossibilities and rank infidelity . . . if God had intended that His creatures travel at the frightful speed of fifteen miles an hour by steam He would clearly have foretold it through His holy prophets." [58] Certainly, the fundamentalist country churches had to fear the effects of increasing urbanism, cosmopolitanism, and worldly sophistication that came to town on the railroad.

An obvious social-psychological result of the railroad was a new consciousness of time. There were two distinct aspects involved in this change: first, an increasing tempo of life that had to be more closely regulated by time; and second, the substitution of standard time for individual or local time.

In prerailroad days businessmen calculated in terms of months. Inventories were replenished by semi-annual trips to wholesale markets, credit terms were six to ten months, or in some rural areas from crop to crop.[59] During the northern winters commercial activity in the interior of the country slowed nearly to a halt. The early nineteenth-century spread of industry and commerce introduced the need for punctuality in employees, but did not speed up the pace of major decision making.

The railroad and the Civil War, nearly contemporaneous in the

[56] *Ibid.*, pp. 71–72.
[57] *Ibid.*, p. 87.
[58] *Ibid.*, p. 249.
[59] Victor H. Clark, *History of Manufactures in the United States* (3 vols., New York: McGraw-Hill, 1929), Vol. II, p. 9; and Lewis E. Atherton, *The Southern Country Store, 1800–1860* (Baton Rouge: Louisiana State University Press, 1949), pp. 52ff.

Middle West, shortened credit terms. The railroad made it possible to do business with much smaller inventories, and the wartime uncertainties and inflation made wholesalers anxious to get their money as soon as possible. Metropolitan salesmen could now visit the shopkeepers and manufacturers of interior towns and sell them orders that could be delivered within weeks, while payment came to be expected within thirty days after delivery.

The increased tempo went deeply into social and business life. Railroad timetables became important in many ways: commutation into big cities grew from the start, children commuted to high school, newspapers and other mail arrived or left by certain trains, visitors came and went by train, and each of these activities had to be set to the minute of scheduled train times.

In the early days of the railroad, time was still read from sundials. While such readings altered only slightly over short distances, between cities a few hundred miles east or west of each other the variation would be several minutes. A railroad junction point, such as Buffalo, would have half a dozen clocks showing the time in other major cities such as Albany, Cleveland, Detroit, or Chicago.

The railroad came to control time, first, by making the station clock the criterion for each town, and second, by establishing four time zones covering all of the United States. The latter change was the result of a decade of timetable conventions, and the ultimate agreement of 1883 superseded fifty-four regional times represented on railroad schedules.[60]

Whatever framework or model used for analysis, almost no one questions the fact that material progress has generally been welcomed in American culture. Robin M. Williams, Jr., in a recent analysis of value orientations in America outlines the major headings of achievement, success, activity, work, efficiency, practicality, progress, and material comfort. These values, all of which would be favorable to the reception of the railroad, make up half of Professor Williams' list.[61] None of the other eight, such as freedom, equality, or external conformity, appear to be of a sort that would oppose acceptance of the railroad. While other students of American culture may define the dominant values or major cultural themes in slightly different words, there seems to be no basic disagreement on the

---

[60] See Robert E. Riegel, "Standard Time in the United States," *American Historical Review*, XXXIII (1927), 84–89.

[61] Robin M. Williams, *American Society* (2nd ed. rev., New York: Alfred Knopf, 1960), pp. 415–470.

general social attitudes of middle-class Americans in the nineteenth century.[62]

Granting that value systems have, by definition, a normative quality, coincidence of the new directions of behavior introduced by an innovation and the sanctioning force of a large part of the value system should produce a strong positive reaction. Put another way, if the changes in social roles required by the innovation are in positively sanctioned directions the role players will respond with high morale and presumably some creativity.[63] These considerations reinforce what has been urged about the innovative effects of the railroad.

If, then, the railroad as an innovation is eagerly received in the culture, it should also reinforce the same traits that account for its reception. By accentuating existing cultural differentials, the railroad presumably made Americans more "American." It seems quite possible that space travel may have similar reinforcing effects.

[62] See Clyde Kluckhohn, "Some Aspects of American National Character," *Human Factors in Military Operations,* ed. Richard Hays Williams (Baltimore: Operations Research Office, Johns Hopkins University, 1954), pp. 118–120; and Thomas C. Cochran, "The Social Scientists," *American Perspectives,* eds. Robert E. Spiller and Eric Larrabee (Cambridge: Harvard University Press, 1961), pp. 110–112.

[63] See Thomas C. Cochran, *The Inner Revolution* (New York: Harper and Row, 1964).

7

Political Impact: A Case Study of a Railroad Monopoly in Mississippi

Robert L. Brandfon

This chapter describes the political repercussions of the Illinois Central Railroad's activities in the State of Mississippi during the latter part of the nineteenth century. At that time the Illinois Central was Mississippi's largest and richest corporation. The national scope of its operations was admirably suited to post-Reconstruction southern visions of a New South forever linked in prosperous harmony with triumphant northern capitalism. But the glowing economic benefits expected from the presence of the Illinois Central failed to materialize. The railroad, a private corporation, had its own objectives. Moreover, what economic benefits did accrue were unevenly divided and served to accentuate long-standing sectional differences within the state. The political impact of this economic situation was a shift in southern thinking about the meaning of ties with northern capital. In the persecution of the Illinois Central for back taxes, described in this paper, the people of Mississippi were not seeking to curb the railroad's privilege of running its own business; and they were not turning their backs on earlier goals of attracting capital (although it is hard to see how persecution would not prevent this). Rather they were attempting to force the Illinois Central to contribute in a more meaningful way to the economic uplift of the state.

During the late nineteenth-century expansion of the American railroads, the South experienced a new northern colonialism. Before the Civil War, the South had built a rough and unco-ordinated railroad system. Although aided by some northern and foreign capital, it was essentially a southern affair, indigenous to southern capital resources and reflecting the South's economic necessities. It was also a sectional rail system with few connections into the North, and with a basic east and west orientation. The Civil War destroyed the physical plant of this system, and, in addition, carried away the capital base that might have resurrected it. Exhausted, physically prostrate, and thoroughly bereft of capital resources, the South's only hope for rebuilding its railroads and its economy lay with the help of northern capital.

During the nineteenth century, the railroads were both promoter and object of capital investment. Aware of the South's financial incapacity, postwar southern leadership actively sought the southern extension of northern rail systems, assuring the voters that this would bring about a new millennium. Northern railroad capital eventually did venture south, but its decisions to do so were not based on southern necessities. Rather, these decisions were geared to northern objectives, without regard for their economic and political impact on the South.[1] The fact that northern railroad expansion was motivated by exigencies beyond southern control, with the objective of putting profits into nonsouthern pockets, lies at the heart of southern reaction to it. The Illinois Central's expansion into Mississippi during the 1870's well illustrates this point.

Its decision to expand south had nothing whatever to do with Mississippi. Indeed, during the height of the railroad's political troubles in that state, the road's president, Stuyvesant Fish, recalled that the board of directors, fearful at the state's prewar reputation of repudiation, studiously avoided including the name of Mississippi in the Illinois Central's new southern extension, even though the major part of it traversed the state.[2] The board's objective was to connect the port of New Orleans with Chicago, thus completing the earlier dream of linking the Gulf states with the Lake states by a solid bond of railroad line. Motivated by the political ambitions of Senator Stephen Douglas of Illinois during the 1840's,[3] this dream took a generation to realize.

In order to breathe life into his Lakes-to-the-Gulf vision, Senator Douglas secured a 2,500,000-acre federal land grant in Illinois for the purpose of encouraging a railroad to run south from Chicago. In 1851, as a result of this largesse, the Illinois Central Railroad came into being. In its efforts to sell its huge land holdings

---

[1] The only volume on southern railroads after the Civil War is John F. Stover, *The Railroads of the South, 1865–1900* (Chapel Hill: University of North Carolina Press, 1955).

[2] Stuyvesant Fish to Jerome Hill, February 15, 1900, *Stuyvesant Fish Letters* (Illinois Central Railroad Archives, Newberry Library, Chicago).

[3] Paul Wallace Gates, *The Illinois Central Railroad and its Colonization Work* (Cambridge: Harvard University Press, 1934), pp. 26–32.

to farmer-settlers and thus build up future agricultural freight, the Illinois Central was, in its early years, more of a land company than a railroad.[4] Nevertheless, in a short time the road extended its tracks from Cairo, Illinois, at the confluence of the Ohio and Mississippi rivers, northward in the form of a Y, one terminal to the northeast of the state, at Chicago, and the other to the northwest, at Dunleith. The location of its track and the extent of its land holdings gave the Illinois Central Railroad a leading role in the maturing economic and political life of the State of Illinois. The road was keenly aware of the importance of its contribution in promoting settlement, and it acted accordingly. By the mid-1860's the Illinois Central basked in the sunshine of prosperity, hauling the huge agricultural outpourings from the interior of Illinois (then the nation's grain basket) into Chicago, from there to be sent eastward via the Great Lakes.[5]

Ironically, the Illinois Central's success at this time was the cause of its near extinction. Competition from eastern trunk lines not only caused the diminution of the Lakes trade, but began seriously to cut into the Illinois Central's trade in the interior of Illinois. By the end of the 1870's, for example, its main line was tapped by rival roads at no less than forty-nine different points.[6] The fierce intensity of this competition, coupled with the panic and depression after 1873, put its future in jeopardy. Harassment by Illinois' "granger laws," and the failure of the wheat crop in 1876, added to the railroad's troubles. Stock values and dividends declined drastically. "The financial distress," wrote President Stuyvesant Fish, "was such at that time that after all the assets in the Company's treasury had been hypothecated some of the directors had to come to the rescue by pledging their individual credit on the Company's notes."[7] It was in these dark times that the youthful Stuyvesant Fish, son of Secretary of State Hamilton Fish, and heavily endowed with wealth and social position, was prevailed upon by the indomitable William H. Osborne to become a member of the Illinois Central's board of directors.

President of the road during the boom period of its growth, Osborne must be credited with saving it from eventual bankruptcy.[8]

[4] *Ibid.*, p. 147.

[5] Carlton J. Corliss, *Main Line of Mid-America, The Story of the Illinois Central* (New York: Creative Age Press, 1950), pp. 81–89.

[6] *The Commercial and Financial Chronicle,* XXIV (February 10, 1877), 134; XXXII (March 19, 1881), 302.

[7] Stuyvesant Fish to Robert W. Patterson, January 17, 1894, *Stuyvesant Fish Letters.*

[8] Corliss, *op. cit.*, pp. 172–173; *The Commercial and Financial Chronicle* XXXII (March 19, 1881), 302.

His solution to the Illinois Central's troubles was a bold one. In the vicious world of cutthroat enterprise, cutting costs was at best a stopgap measure retarding, but not turning away, inevitable failure. Only by expanding trade to increase earning power could a railroad survive its competition. No one would dispute these observations, but Osborne's frontier for expansion seemed a desperate gamble. Pinning his hopes on a revival of the port of New Orleans, which was suffering from the ill effects of a persistent sand bar at the entrance way to the Mississippi, Osborne proposed to extend the Illinois Central south from Cairo through the prostrated states of Kentucky, Tennessee, and Mississippi. The board of directors, upon an investigation, gave Osborne a green light. After cautious experimentation with two local southern railroads, the board decided to take them over and consolidate them. In 1877, the Chicago, St. Louis and New Orleans Railroad was incorporated, effecting the first and only through line from New Orleans to Chicago. Success was immediate and was stimulated further by the successful clearing of the Mississippi's mouth through the engineering genius of James B. Eads,[9] and by the return of national prosperity beginning in 1878. Within the short space of three years the value of traffic on the Illinois Central's southern division from New Orleans to Cairo increased sixfold. Declared by the board in January 1881 to be "an absolute success," [10] the extension, in Fish's words, "proved to be the salvation of the property." [11]

Adding to the general glow of good feeling among railroad officials was the enthusiasm with which Mississippi's political order greeted the southern extension. The entrance of such a powerful northern railroad corporation was hailed as the beginning of a new era in Mississippi's history, reflecting the all-essential confidence of northern and foreign capitalists in the state's future. A firm belief in the efficacy of a national railroad automatically to create prosperity had become a shibboleth. The specific forms of its economic contributions, however, were never clearly defined. Minds not clouded by the glowing rhetoric of the Atlanta *Constitution*'s in-

[9] Florence Dorsey, *Road to the Sea: The Story of James B. Eads and the Mississippi River* (New York: Rinehart, 1947), pp. 166–217; E. L. Corthell, *A History of the Jetties at the Mouth of the Mississippi River* (New York, 1880), pp. 60–68, 165–174, 224–238; *New York Tribune*, February 11, 25, March 27, April 3, 1879.

[10] Illinois Central Railroad, *Report of the Directors*, February 1, 1877, and January 22, 1881 (Illinois Central Railroad Archives, Newberry Library, Chicago).

[11] Stuyvesant Fish to Robert W. Patterson, January 17, 1894, *Stuyvesant Fish Letters*.

trepid editor, Henry Grady, would have seen that the Illinois Central's sole interest in entering the State of Mississippi was not for any developmental purpose but to reach New Orleans by the shortest possible route. Two other facts were also evident. First, the southern extension ran the mid-section of the length of the state. This area of scrub pine and poor soil was in permanent agricultural decline; any hopes of reaping great amounts of freight traffic along the railroad's right-of-way were minimal. Second, unlike its experience in Illinois, the Illinois Central did not have a land grant in Mississippi and thus lacked the power, or the interest, to promote agricultural settlement. The belief that the mere presence of the Illinois Central would bring prosperity to the state, or that the railroad would actively bring it about, was an illusion.

It was equally illusory to assume that the Illinois Central Railroad could possibly reshape economic trends within the state. By the 1880's, Mississippi's economic outlines had already taken shape. Lacking mineral resources, this was an agricultural region specializing in cotton growing. The older areas of cotton specialization, such as the black prairie belt in the eastern part of the state and the area around Natchez, however, were losing their once great capacities for profitable yields. A new center of cotton growing was replacing them. This was the Yazoo Mississippi Delta, an area earlier scorned as an unhealthy disease-ridden swamp. The Delta was the dumping ground for periodic overflows of the Mississippi River's silt deposits gathered from the river's continental watershed. Untold centuries of flooding and of receding waters had built up the soils of the Yazoo Delta, piling up layer upon layer of yellow-brown alluvium, and forming in its undrained pools of swamp and bog some of the richest soil in the United States.[12] Lying in the northwestern corner of the State of Mississippi, the Yazoo Mississippi Delta extended southward from Memphis. Its western border was the Mississippi River; its eastern boundary was formed by a north-and-south line of high bluffs and by the Yazoo River, which joined the Mississippi at Vicksburg. The combination of these boundaries, and the alluvial nature of its soil, made the Yazoo Mississippi Delta more than just another frontier. It presented a challenge to reclaim great agricultural riches from the swamp forsaken by earlier pioneers.

A number of developments, occurring chiefly after the Civil War,

[12] For a classic account of the Mississippi River and its action upon the Yazoo Delta, see U.S. Congress, *Reports on the Ohio and Mississippi Rivers made by . . . Charles Ellet, Jr., in 1852*, 63rd Cong., 2nd Sess. (1914), House Document 918.

worked to overcome this new and strange wilderness and promote agricultural settlement. The first of these was the attractiveness of the Delta's hardwood forests. These became an El Dorado for a host of land speculators seeking to cash in on new demands for southern timber. Second, through the creation in 1879 of the Federal Mississippi River Commission, the Delta was assured of national support in replacing and strengthening its system of levees, destroyed during the war by the combined efforts of the Mississippi River and the armies of Ulysses S. Grant.[13] Beginning in 1873, the U.S. Army Engineers, on the pretext of clearing the hulks of sunken ships, worked to clear the ever-forming snags and other obstacles from the Delta's main streams. This assured easy river-boat communication with market outlets, as well as helping to promote land drainage, which was vital to new settlement.[14] With the rolling back of the forest and the draining of the swamps, increasing numbers of Negroes entered the area. Pushed off the marginal and worn-out lands of the older South, this cheap labor force was attracted to the broad and fertile flatlands of the Yazoo Delta where higher yields would theoretically alleviate the crushing burdens of tenant poverty.[15] As for the Delta's white settlers, the natural advantages of river transportation, land, and labor, quickly elevated them from farmer to planter, thus gaining for them the economic and social appurtenances associated with this highest of southern traditions. Although its four million acres were barely cultivated by the early 1880's, the Yazoo Delta already symbolized new hope for Mississippi's economic future. At the same time, however, the bluffs and the Yazoo River, separating the Delta from the rest of the state, became more than a mere geographical cleavage between Delta and hill lands. They became a barrier between planter and redneck, rich and poor, blessed and despised.

With the introduction of railroads into the Delta early in the 1880's, these divisions were intensified. The efficient and reliable railroad transportation easily replaced river-boat traffic and stimulated the latent urges of the planter to increase production and

[13] These developments are traced in Arthur DeWitt Frank, *The Development of the Federal Program of Flood Control on the Mississippi River* (New York: Columbia University Press, 1930); Stanley F. Horn, *This Fascinating Lumber Business* (Indianapolis: Bobbs-Merrill, 1943); Nollie Hickman, *Mississippi Harvest; Lumbering in the Longleaf Pine Belt, 1840–1915* (University, Miss.: University of Mississippi Press, 1962).

[14] This development can be traced only in the U.S. Army Engineers, *Annual Reports*, 1873–1919

[15] See Vernon L. Wharton, *The Negro in Mississippi, 1865–1890* (Chapel Hill: University of North Carolina Press, 1947), pp. 106–116.

maximize profits. While the lines of the Illinois Central Railroad in central Mississippi ran "simply as a bridge through a desert,"[16] another national railroad network began to exploit the Delta's potential for rich harvests of cotton freight. The Louisville, New Orleans and Texas Railroad was the creation of one of America's great railroad entrepreneurs, Collis P. Huntington. It was formed in 1884 as the rail bridge joining the Chesapeake and Ohio Railroad at Memphis to the Southern Pacific at New Orleans, thus forming a national, super trunk line stretching in a huge arc from Portland, Oregon, to Newport News, Virginia.[17] Although this colossus had a brief life, by the end of the decade, the Louisville, New Orleans and Texas Railroad had joined Memphis with New Orleans by a through trunk line. In addition it had built up a network of over four hundred miles of track in the Yazoo Mississippi Delta, placing that area's transportation firmly on a north-and-south axis and giving the products of its soils major market outlets. When the Huntington empire collapsed as a result of its founder's overexpansion, the Louisville, New Orleans and Texas Railroad continued to thrive, chiefly because of the huge outpouring of agricultural freight from the Yazoo Delta. Indeed, in 1889, the annual report complained of the railroad's inability to handle it all.[18]

The successful exploitation of the Yazoo Delta trade by the Louisville, New Orleans and Texas Railroad revealed to the directors of the Illinois Central the first threat to their assumed hegemony over north-and-south rail traffic out of New Orleans. The directors could brook no serious interference with their southern extension, the lifeline of their railroad. In the first place, the Louisville, New Orleans and Texas Railroad had cheaper costs, made possible by the flat contour of the Delta lands. This allowed each locomotive to haul double the normal number of freight cars carried over the steep grades of the Illinois Central's main line through central Mississippi. Second, the Louisville, New Orleans and Texas had inherited a 700,000-acre land grant in the Delta, giving it right-of-way advantages, as well as certain control over agricultural settlement. To the directors of the Illinois Central, the importance of these benefits was not unfamiliar: they were decisive factors in the growth of their railroad in Illinois.

[16] Stuyvesant Fish to Jacob M. Dickinson, November 26, 1900, *Stuyvesant Fish Letters*.

[17] Stuyvesant Fish to Collis P. Huntington, July 13, 1888, *ibid.*; Cerinda W. Evans, *Collis Potter Huntington* (Newport News, Va.: Mariners' Museum, 1954), Vol. II, pp. 576–579.

[18] Louisville, New Orleans and Texas Railroad, *Annual Report*, 1889.

Earlier in the decade of the 1880's, the Illinois Central had toyed with the idea of extending a branch line northwestward from the state capital at Jackson to some point on the Mississippi River, thereby cutting across the Yazoo Mississippi Delta. For this purpose they had received a liberal charter from the state legislature, eager to promote every kind of railroad development.[19] Thus was born the Yazoo and Mississippi Valley Railroad, but it was thwarted by the zealous activities of the Louisville, New Orleans and Texas, and was unable to draw off much of the Delta's rich agricultural freight. Increasingly nervous, the directors of the Illinois Central leapt at the opportunity, in 1892, to buy out the Louisville, New Orleans and Texas, the financial structure of which had been undermined by Huntington's ambitions elsewhere. As a result of the purchase, the Illinois Central, a blue-chip stock at this time, owned the majority of the railroad mileage in Mississippi and, in addition, was the largest owner of the richest lands in the state. Renamed the Yazoo and Mississippi Valley Railroad, the Louisville, New Orleans and Texas was forever erased from the memory of Mississippi's citizenry who now contemplated this large northern corporate power in their midst and questioned the effects it would have upon their future.

This questioning of the Illinois Central's rich power reflected a re-examination of the basic assumptions about the sanguinary influence of northern capital: indeed, whether that capital was perhaps a force for evil rather than for good. Prolonged years of agricultural decline motivated this thinking. Outside the Yazoo Delta, economic conditions for most of Mississippi's farmers had never returned to the normalcy associated with ante bellum days. Their naturally poor soils were continually depleted. This condition was caused in large part by ignorance of good farming methods as well as by intensive cultivation, made necessary by declining cotton prices. For the yeoman white farmer, tenancy was the first step downward to the peonage of the Negro. With every prospect toward tenancy, agricultural fundamentalism rose in an inverse ratio. The basic tenet that the yeoman farmer Democrat was bound to prosper, that his prosperity was the backbone of national strength, this faith was given renewed amplification. Jeremiads gave way to the anger and suspicion inherent in fundamentalist judgment. Only an outside force, a conspiracy, they thought, could

[19] *Mississippi Session Laws, 1882*, pp. 838–849; J. C. Clarke to W. K. Ackerman, November 22, 1881, and January 19, 1882, *Ackerman Letters* (Illinois Central Railroad Archives, Newberry Library, Chicago).

have brought about the godless perversion that had reduced Mississippi's yeomen to the bottom rung. The finger of accusation pointed beyond the Delta planters, who had not suffered, to the tools of their deliverance — the Negro and the railroad corporation. These beliefs came alive in the mid-1880's and swept through the State of Mississippi kindling flames that were to burn the entire southland and cause the most significant political revolution in that section's history.

Attacks upon the corporation followed the ambivalent course of political opportunism. No one sanctioned abolition of the railroad; there was, however, general agreement that its power should be proscribed. What exactly was its power? The answer was quickly forthcoming: privilege. Devastated by war, without credit or cash, and possessing lands whose worth was unrecognized or had depreciated, Mississippi's reconstruction governments had sought to attract outside capital investment by granting liberal tax privileges. For railroads, the standard form, written into legislative charter grants, was the privilege of creating a thirty-year reserve fund composed of all taxes the railroad should have paid to the state. From this fund, the railroad could pay its construction costs and other initial liabilities. If, however, railroad net profits allowed it to pay annual dividends of more than 8 per cent above its debts and liabilities, the tax exemption would cease. Counties and municipalities followed the state's example by offering other forms of tax exemptions.[20] Although these privileges were not granted to the Illinois Central's main line (the Chicago, St. Louis and New Orleans Railroad), they allegedly were granted to the Louisville, New Orleans and Texas Railroad and were included in the transfer of its property to the Yazoo and Mississippi Valley Railroad. In this way, more than half of the Illinois Central's system in Mississippi was tax exempt. For one of the nation's most stable and wealthiest railroads to insist upon the continuation of this bonanza while operating in one of the nation's poorest states was an invitation to political assault.

The attacks were staged initially in the local courts, or more specifically, the jury-trial courtroom. In rural areas, the jury courtroom is not merely a place where facts are determined. As the arena of community gossip, it is an emporium for legal horsetrading

[20] *Mississippi Session Laws, 1870,* pp. 269–327; Edward Mayes to Robert T. Wilson, March 13, 1895, *Stuyvesant Fish Letters;* Hunter C. Leake (compiler), *Illinois Central and Yazoo and Mississippi Valley Railroad Company vs. Wirt Adams, Briefs* (Mississippi Department of Archives and History, Jackson, Miss., 1908), pp. 9–21, 30–35.

and a made-to-order political stump for aspiring country lawyers. In this setting, the Illinois Central was a natural target. It was a foreign corporation and it was rich, an all too tempting combination for a legal brotherhood made lean by accumulating years of general economic decline. As one railroad lawyer explained it, the proliferation of personal injury suits against the Illinois Central in Mississippi was caused by "the general shortage of pasturage for lawyers in any other field of litigation." [21] His reasoning was supported by statistics. In 1900, for example, no less than half of the Illinois Central's 1,193 legal cases were in Mississippi. This situation did more than offer a living to Mississippi's lawyers. In hot pursuit of a lucrative judgment, it was too easy to set the alien railroad leviathan against the helplessly injured railroad employee or the poor farmer whose only mule had ill-advisedly strayed across an unfenced railroad track. The juries, stirred to the proper pitch of righteous indignation by the impassioned harangues of tomorrow's legislators, made their awards worth the candle. Hostile local sentiment built up in this way against the railroad was inadvertently aggravated by the railroad's policy of promoting the most judicious members of Mississippi's bar to high salaried positions with the Illinois Central.[22] With few other opportunities left to them, those not so annointed with lucrative railroad positions were left in the jury courtroom, there to vent personal resentments against the Illinois Central upon primed ears eager to hear more.

State government policy reflected local sentiment. Here too, hostile attitudes toward the rich railroad corporation were motivated by economic considerations. At the outset of the 1870's, the New South leadership based its demands for home rule and the Democratic party on the alleged waste and corruption of Black Republicanism. The return to the "wise and frugal administration" of the Democratic party in 1875 signaled drastic reductions in state social services with a corresponding cut in taxation, especially on lands, the base of the state's tax structure. The brief return in 1878 of a modicum of prosperity and an accompanying wave of optimism about the future, raised irresistible demands, particularly by Yazoo Delta planters, for more roads, bridges, courthouses, and other capital improvements.

Wishing to meet these demands, yet fearing to open themselves to the same charges so recently leveled against Reconstruction

[21] Edward Mayes to Jacob M. Dickinson, November 1, 1901, *Stuyvesant Fish Letters.*
[22] Robert C. Beckett to Anselm J. McLaurin, November 24, 1896, *McLaurin Papers* (Mississippi Department of Archives and History, Jackson, Miss.).

leadership, Mississippi's New South government resorted to floating bond issues rather than making any distasteful increases in taxation. The fruit of this policy, however, was increasing treasury indebtedness — by 1895 reckoned at over $3,000,000. Persisting for more than a decade, indebtedness threatened seriously to undermine the state's credit (already tarnished by its prewar repudiation of state bank bonds), and made new bond issues impossible. Moreover, the spectre of faster growing indebtedness became ominous as a result of widespread declines in agricultural prices. By the mid-nineties, the worst depression in history had settled, without exception, upon every corner of the state.

In desperation, Mississippi's state government searched for ways to raise additional revenue. In January 1894, the governor urged the authorization of another bond issue and a general increase in taxation. Neither of these measures was acceptable to legislators harassed as they were by the economic laments of their constituents. An abortive attempt was made to issue state warrants as legal tender. Because of their similarity to federal notes, these warrants were confiscated and the state's governor, treasurer, and auditor were momentarily arrested by secret service agents on charges of counterfeiting.[23] Almost as a last resort, the administration, with the full support of the legislature, decided to undermine tax privileges previously granted to many out-of-state corporations, in particular to the richest of them all, the Illinois Central's Yazoo and Mississippi Valley Railroad. Although general sentiment favored outright repeal of these privileges, it could not be done so simply. Tax exemptions, which ran for a specified term of years, were written into corporation charters. It was, therefore, expedient to prove that the corporation had violated the conditions for its tax privileges. In this way, the corporation would owe the state and local governments back taxes as well as surrendering tax privileges for the remainder of the exemption period. Since jury courtroom feelings ran high against the Illinois Central in particular and against all corporations in general, it was not expected that proof would be difficult to find.

The burden of proof fell upon the state revenue agent, Wirt Adams, Jr., whose office was given new legal powers by the legislature to go after tax privileges. Personal incentives were added.

[23] Governor's Message quoted in Mississippi *House Journal, 1896*, pp. 17–21; J. G. Carlisle to U.S. Attorney General, August 18, 1894, *Justice Division File*, pp. 9149–9194 (U.S. National Archives, Washington, D.C.).

Whatever the outcome of his legal activities, administrative and court costs would be borne by the state. His salary and those of his staff would be paid solely from the 20 per cent commission on all collections of back taxes received from delinquent taxpayers. No longer appointive, his office was now made elective. His tenure would thus be dependent upon how much he could gather. Adams succeeded beyond all expectations. At the time of his death in 1914, after twenty-eight consecutive years in office, he was responsible for wresting from numerous corporations in Mississippi an estimated $7,000,000 of additional tax revenue for state and local governments.[24] His largest haul was the judgment he received against the Illinois Central. As a result of ten years of persistent legal and political efforts, he gathered from the railroad $1,500,000 in back taxes, and he forced it to forfeit an additional $1,000,000 in future tax exemptions.

His course against the Illinois Central was as persistent as his arguments were simple. First, he argued, the underhand methods used by the Louisville, New Orleans and Texas Railroad in watering its stock and passing its profits to a "secret ring of alien financiers" in New York made it virtually certain that the railroad's net profits would never exceed 8 per cent above its debts and liabilities. In this way, the road had illegally maintained its tax privileges at the expense of Mississippi's taxpayers. Second, Adams charged, the Louisville, New Orleans and Texas Railroad had competed with the Illinois Central for the north-south trade out of New Orleans. Under Mississippi law it was illegal for competing railroads to combine. The Illinois Central had tried to circumvent this restriction by having its tool, the Yazoo and Mississippi Valley Railroad, buy out the Louisville, New Orleans and Texas Railroad while all the time maintaining the fiction that the Yazoo and Mississippi Valley was independent. Long before the sale was made official, Adams continued, members of the Illinois Central board of directors had composed the secret ring of New York financiers who were reaping the profits of the Louisville, New Orleans and Texas. The circumstances of the sale, therefore, were "a sham and a fraud" designed to avoid the law and to claim undeserved tax privileges. Lastly, as a result of the sale, an entirely new railroad was created, thus abrogating any prior tax exemptions claimed

[24] *Mississippi Register, 1904* (Jackson, Miss., 1904), J. B. Harris to Stuyvesant Fish, September 2, 1897, *Stuyvesant Fish Letters;* Jackson, Miss., *Clarion-Ledger,* April 26, 1914.

by the Louisville, New Orleans and Texas. All these arguments were calculated to appeal to widespread fears about the conspiratorial and alien qualities of the Illinois Central.[25]

The railroad's tortuous reply to these charges served only to arouse more popular suspicion against it. The Illinois Central denied owning the Louisville, New Orleans and Texas Railroad before its consolidation with the Yazoo and Mississippi Valley. The only connection between the two roads was through the Mississippi Valley Company, a sort of holding company chartered by the Mississippi legislature in 1872. The Illinois Central owned some stock of this holding company, as did some of the members of its board of directors. The Mississippi Valley Company, railroad lawyers argued, and not the Illinois Central, bought the stock of the Louisville, New Orleans and Texas. This stock was then transferred to the United States Trust Company in New York in order to secure some bonds of the Illinois Central. Thus, there was no connection between the Illinois Central and the Louisville, New Orleans and Texas Railroad before or after its consolidation with the Yazoo and Mississippi Valley. Because it was a consolidation rather than a purchase, the Yazoo and Mississippi Valley had every right to the tax privileges possessed by the other. Railroad lawyers substantiated this argument by pointing vigorously to several Mississippi Supreme Court decisions upholding the state's right to grant tax exemptions as well as the right of the Yazoo and Mississippi Valley to inherit them.[26] These tax privileges, railroad attorneys warned, amounted to a solemn, binding, irrepealable contract between the railroad and the State of Mississippi. To abrogate this contract by attacking the Yazoo and Mississippi Valley's clear and indisputable right to its tax exemptions would blot the State of Mississippi with a fresh stain of repudiation.

This dire warning had a limited effect. A jury decision in 1896 to determine whether the railroad had been able to grant dividends of more than 8 per cent favored a compromise, granting some back taxes to the revenue agent and allowing the railroad to keep the rest. The verdict was appealed to Mississippi's Supreme Court

[25] See "State of Mississippi vs. Yazoo and Mississippi Valley Railroad," in Hunter C. Leake, *Briefs;* Edward Mayes and J. B. Harris to James Fentress, February 17, 1894, *Stuyvesant Fish Letters;* Edward Mayes and J. B. Harris to Stuyvesant Fish, May 14, 1895, *ibid.*

[26] *Mississippi Mills* vs. *Cook,* 56 *Miss.* 40; *McCullock* vs. *Stone,* 64 *Miss.* 378; *Yazoo and Mississippi Valley Railroad* vs. *Thomas,* 65 *Miss.* 553; *Natchez, Jackson and Columbus Railroad* vs. *Lambert,* 70 Miss. 779: James Fentress to Stuyvesant Fish, June 21, 1898, *Stuyvesant Fish Letters;* James Fentress to Yerger and Percy, November 29, 1892, *ibid.*

because both sides sought a clear-cut decision: the railroad, because agreement to compromise would jeopardize its future tax exemptions, which would expire in 1907; the revenue agent, because he was certain he could not only collect all back taxes for the past ten years, but could also force the railroad to forfeit its future exemptions. Adams' decision was based on an informed conviction that popular trends against the Illinois Central and the principle of tax exemptions in general would soon be given practical political form. The legislative session in January 1896 had seen the introduction of a bill "to repeal all laws exempting railroads from taxes in the State of Mississippi." [27] Although this bill was defeated in a house committee, the strong pressure put on the committee members by the Illinois Central was resented, and opened the railroad to charges of meddling in politics. Despite the bill's defeat the revenue agent's cause was given new support by the fact that the bill was introduced and supported by citizens of the Yazoo Delta, who by all rights should have favored the railroad.

More than any other people in the state, the planters of the Yazoo Delta had benefited from the presence of the railroad. The aggressive activities of the Louisville, New Orleans and Texas Railroad, crisscrossing every important growing area, had driven away inefficient and unreliable river-boat traffic and stimulated intense and highly profitable cotton cultivation in undeveloped parts of the Delta. After 1892, the Yazoo and Mississippi Valley Railroad, eager to monopolize the large profits of its predecessor, succeeded in choking off every potential rival, including the mighty Southern Railroad, whose competition in the Yazoo Delta never amounted to more than a mere shadow. Drastically reduced cotton prices, beginning in 1891, led to a curtailment of cotton expansion. With the decline came a greater awareness of the importance of competition in lowering freight costs. Planters were agreed that the Yazoo and Mississippi Valley Railroad was certainly more efficient than the river boats had ever been, but they bemoaned the loss of competition. Moreover, the Yazoo and Mississippi Valley Railroad had not tried, they thought, to encourage the expansion of cultivation by extending its lines into the Delta's interior or by using its vast financial resources to uplift the planter community upon which it fed. To many planters, the railroad was only willing to reap the profits of the harvest but not to sow them.

[27] Mississippi *House Journal, 1896,* p. 396; *Jackson (Miss.) Evening News,* February 3, 1896.

These feelings gained wider support during the spring of 1897 when the greatest flood in the Delta's history up to that time caused widespread devastation. This was followed by a further decline in cotton prices; by early 1898 they reached their lowest level for the nineteenth century. Even if it had not been rumored that the Yazoo and Mississippi Valley was attempting to do away completely with the Southern Railroad's lines in the Delta, the combination of flood disaster and continually falling prices would have brought about some adverse reaction. In the 1898 session of the legislature, the Delta planters were instrumental in enacting a bill that forbade competing railroads operating within twenty miles of each other to purchase or lease an opposing line.[28] Since this was not as direct an attack as was the measure to withdraw tax exemptions, passage was easy. The restriction prevented the Yazoo and Mississippi Valley's attempt to swallow up the Southern Railroad's Delta lines, but did nothing to dampen its desire to control every other part of the Delta's rail traffic: with the return of prosperity in 1899, the Illinois Central Railroad, during the next half dozen years, extended its Delta rail network by 33 per cent, even buying up the tiny narrow-gauge logging roads that appeared from time to time.

The establishment of a virtual monopoly within the Yazoo Delta, by then Mississippi's most productive region, presented the Illinois Central Railroad with an entirely different set of responsibilities toward the state. The southern extension during the late 1870's had had as its object a link between the cities of Chicago and New Orleans, the State of Mississippi being merely a passageway. Twenty years later, the Illinois Central had an additional objective — the exploitation of the rich trade of Mississippi's Yazoo Delta. In this new acquisition, the railroad's role had shifted from that of transient boarder to permanent resident, a rich and influential resident with a broad set of developmental responsibilities like the ones it possessed in Illinois. Illinois Central officialdom was not wholly cognizant of the significance of this shift in roles, and seemed to lack the required perceptiveness and sensitivity to the differing problems of responsibility in Mississippi. On the one hand, President Stuyvesant Fish could assert that his railroad was more interested in Mississippi's development than any other person or corporation within the state, adding that "everything affecting the State for good or evil affects us in like manner, and this for all

[28] Mississippi *Senate Journal, 1898*, pp. 43, 79; Mississippi *House Journal, 1898*, p. 175; *Mississippi Session Laws, 1898*, pp. 95–96.

time." [29] On the other hand, the interests of his railroad and the satisfaction of his stockholders were paramount over any other interests. This, according to Fish, was what was best for the public, which had to understand "that anything which detracts from the profitableness and efficiency of the railways injuriously affects all other interests in our common country. . . . The magnitude of the interests involved are too great to allow of any other course." [30]

Unconscious of its role as Mississippi's richest corporation, the Illinois Central never clearly defined its own position in the midst of a poor agricultural state with no industrial potential. Neither was its position rationally defined by the people of the State of Mississippi. In the midst of desperately hard times, resort was had to harassment by the state, intimidation by the railroad, and alienation for both.

Further fruits of this condition were revealed, in June 1898, by Mississippi's Supreme Court appellate ruling on the verdict of the jury trial. The state high court upheld the revenue agent's position without exception, finding the railroad liable to all back taxes and denying its claims to future exemptions. The decision was a blow to railroad lawyers, who had over long years painstakingly built up a legal bastion of precedent to defend themselves from Adams' assaults.[31] It was not, however, unexpected. The Illinois Central lawyers in Mississippi were in tune with the political situation by which the high court was ruled. With popular sentiment against privilege being translated into political action, they warned, it was only a matter of time before the courts too would fall into line.

President Fish could not believe that Mississippi's officialdom, especially the members of the state's highest court, could be swayed by fickle and irrational public feeling. He was convinced that the decision was unfairly motivated by political demagoguery, and that it was equally interlaced with Adams' personal pecuniary benefit.[32] He believed that a solemn appeal to the Supreme Court of the United States, then deeply concerned with the sanctified nature of contractual obligations, would right the wrongs done to his railroad. His advisers were less sanguine. Whatever the outcome of a federal court ruling, charges of stock watering by the Louisville,

[29] Stuyvesant Fish to Jerome Hill, February 15, 1900, *Stuyvesant Fish Letters.*
[30] Stuyvesant Fish to Robert W. Patterson, January 17, 1894, *ibid.*
[31] *Wirt Adams* vs. *Yazoo and Mississippi Valley Railroad,* 75 Miss. 275; Stuyvesant Fish to Sidney F. Andrews, March 8, 1897, *Stuyvesant Fish Letters;* James Fentress to Mayes and Harris, December 28, 1897, *ibid.; Yazoo and Mississippi Valley Railroad* vs. *Wirt Adams,* 77 Miss. 206.
[32] Stuyvesant Fish to James Fentress, July 5, 13, August 15, December 5, 7, 1898. *Stuyvesant Fish Letters.*

New Orleans and Texas Railroad and the related questions of 8 per cent dividends would still have to be determined by a hostile local jury, sure to be angered further by appeals outside the jurisdiction of Mississippi. Only by compromise, Fish's advisers warned him, would anything be saved at all. Under Mississippi law, legislation for this sort of a compromise was mandatory. But it was dangerous to throw all the involved and delicate questions into the political maelstrom of a southern legislature. However, Fish consented to this strategy, at the same time holding in readiness his appeal to the Supreme Court. This could be used if Mississippi's legislature should prove recalcitrant over the railroad's compromise proposal.

Fish's compromise proposal to the legislative session of 1900 was a niggardly one, offering half the total sum being contested.[33] Moreover, a worse time could not have been chosen for its presentation. Throughout the summer of 1899, the issue of the Illinois Central's tax exemptions had been the focus for a heated political campaign. Aspiring to a seat in the United States Senate, outgoing Governor Anselm J. McLaurin had made the issue simple enough. "I want the Illinois Central Railroad Company and the Yazoo and Mississippi Valley Railroad Company to pay their taxes as we pay ours. . . . It is the duty of the State . . . to see that the rich and powerful bear their lawful and just share of the expenses of the government, and that they be not permitted to shift the burden of their taxes onto the poor and the weak." [34] These sentiments were carried from the political stump to the legislative chamber as evidenced by the unwillingness of any member of the lower house to champion the railroad's compromise bill. In the senate, where the railroad lobby was stronger, enough support was found in the judiciary committee to assure a favorable majority report. Attached, however, was a vituperative minority dissent. This charged the railroad with evading its responsibilities and seeking favoritism at the expense of the less powerful. The minority report more accurately reflected the feelings of the entire senatorial body, as the compromise bill was defeated in the final voting by a decisive two-thirds majority.[35] The appeal to the Supreme Court met a similar fate as

[33] Stuyvesant Fish to Jacob M. Dickinson, December 20, 1899, *ibid*.

[34] *Memphis Commercial Appeal*, December 13, 1898; *Governor's Messages* (McLaurin), *1898* (Mississippi Department of Archives and History, Jackson, Miss.).

[35] Mississippi *Senate Journal*, 1900, pp. 295, 318, 452–459, 521–523; Telegram, Stuyvesant Fish to Jacob M. Dickinson, March 8, 1900, *Stuyvesant Fish Letters*.

the federal judges upheld the cause of the Mississippi revenue agent and added as an aside (curious in the days when the Supreme Court, by its staunch defense of contractual obligations, was noted for rulings which meddled with state policy), "The legislature is the proper guardian of the public faith, and . . . whatever policy the State may choose to adopt with respect to encouraging or discouraging the investment of capital from abroad, the duty of the courts is to declare the law as they find it, and avoid the discussion of questions of policy, which are clearly beyond their province. Certainly this court is not the keeper of the State's conscience." [36]

The combination of these reversals forced the Illinois Central to surrender a fortune in back taxes and future exemptions. Large as it was, the total loss of $2,500,000 was hardly felt by the Yazoo and Mississippi Valley Railroad, whose gross receipts, taken largely from Yazoo Delta trade, in any single year after 1900 amounted to more than double its total taxes for twenty years. The outcome of the back tax episode, therefore, caused little if any financial pain. It revealed, however, the utter political weakness of the state's only capitalist giant. Political impotency was new to the long and influential history of the Illinois Central Railroad, whose officials explained away their helplessness in Mississippi as the result of demagogic persecution peculiar to the naturally violent, southern temperament. This was, however, a description of symptoms, not an explanation of causes. The people of Mississippi were united in the harassment of their greatest potential benefactor for the reason that it had remained only a potential. Despite long years of economic adversity, southerners never lost their old faith in the efficacy of northern capital to put their economy upon a sound prosperous basis. Rather, the southern voter was frustrated by northern capital's unwillingness to utilize that power. The objective of the Illinois Central had been to exploit rather than to contribute. As a result, it remained aloof and politically alienated from every interest with which it could naturally have allied. The revenue agent, therefore, received unanimous support in his attempts to force the Illinois Central to contribute something. Since the nature of northern capital's contribution was always loosely defined, the simplest contribution was tax money. Indeed, in the light of the Illinois Central's tremendous earnings from the Yazoo Delta, credit must be given to the state's restraint in taking relatively little. Southern populism and the progressivism that followed need to be

[36] *Yazoo and Mississippi Valley Railroad* vs. *Wirt Adams*, 180 U.S. 25.

viewed in this light: the marrow of these reform movements was not to curb the capitalist giant but to force it to give more.

Possible Parallels

With the beginning of construction on the Mississippi Test Facility in 1963, the federal National Aeronautics and Space Administration has undertaken a task somewhat parallel to the role of the Illinois Central in Mississippi during the late nineteenth century. The proportions of NASA's involvement, however, are grander. Located along the Gulf Coast east of the Pearl River and forty miles from the Michaud base in New Orleans, the test facility, scheduled for completion in 1966, is to cover a core of 13,500 acres surrounded by a 128,000-acre buffer zone. Plans include the improvement of fifteen miles of river channels and the construction of an additional fifteen miles of canals for purposes of transporting bulky equipment and fuels that cannot be transported by rail. Total construction costs have been estimated at $111,000,000.[37] In the construction of this site for rocket testing, the federal government has guaranteed a huge financial expenditure to fulfill an object that has no relation whatever to the indigenous needs of the State of Mississippi. Again, as with the Illinois Central, the federal government's huge capital expenditure is a definite commitment to the state from which there can be no withdrawal.

The federal government, like Stuyvesant Fish sixty years earlier, has recognized the permanency of its commitment. Recently, the director of the Marshall Space Flight Center at Huntsville, Alabama, Dr. Wernher Von Braun, told his audience in Bay St. Louis, Mississippi, that "We are here to stay and to grow." The sentiment was cheered enthusiastically, but its meaning for Mississippi has yet to be defined. Dr. Von Braun and his associates emphasized, for example, the need for a broader and improved educational system for the communities surrounding the Mississippi Test Facility site. The purpose is to draw upon local talent rather than calling upon out-of-state help. In addition, the over-all uplifting of educational facilities, it is reasoned, will help to attract other industries into the state.[38] Thus, like the Illinois Central the monied giant of an earlier period, the federal government appears now as the

[37] U.S. Congress, Senate, Committee on Aeronautical and Space Sciences, Hearing on S. 1245, 88th Cong., 1st Sess. (1963), p. 2; *Hancock County Eagle*, December 12, 1963.

[38] *The Sea Coast Echo*, May 14, 1964.

new progenitor of industrial prosperity and social uplift. Will this prove a repetitive mirage, a mere shibboleth? Or will it be possible for the federal government to project for itself a positive, progressive, active role within the State of Mississippi? What contribution, for example, will the federal government make toward uplifting the general level of education within the state so as to draw more readily upon local talent? Can the federal government play such a role on a racially segregated basis? If the nation's greatest capital source concentrates upon its primary object of testing rockets in the nation's poorest state where there are no other large sources of capital, it might soon find itself resented and suspected. The charge of exploitation without contribution (colonialism, if you will) might be repeated and intensified among people whose state's rights feelings are an almost immutable tradition. Without active participation in the life of the state, the federal government might find itself alone against an array of hostile local forces.

8

The Impact of the Railroad on the American Imagination, as a Possible Comparison for the Space Impact

Leo Marx

From a period long before the Christian Era down to 1829 there had been no essential change in the system of internal communication. At present, before another half century has elapsed, the Cumberland Turnpike is as antiquated as the Appian Way, — as useful, perhaps, but far less interesting.

As to the railroad system, it long ago became impossible to compute the number of miles contained in it or the millions of capital which its construction had cost. . . .

Though this material or financial aspect of the system is that which is almost invariably dwelt upon, it is by no means the most interesting one. Here is an enormous, an incalculable force practically let loose suddenly upon mankind; exercising all sorts of influences, social, moral, and political; precipitating upon us novel problems which demand immediate solution; banishing the old before the new is half matured to replace it; bringing the nations into close contact before yet the antipathies of race have begun to be eradicated; giving us a history full of changing fortunes and rich in dramatic episodes. Yet, with the curious hardness of a material age, we rarely regard this new power otherwise than as a money-getting and time-saving machine. We know sufficiently well the number of passengers and of tons of freight which the railroad system annually moves; . . . but not many . . . ever stop to think of it as, with perhaps two exceptions, the most tremendous and far-reaching engine of social change which has ever either blessed or cursed mankind.

Charles F. Adams, Jr.,
"The Railroad System," 1868.[1]

[1] "The Railroad System" was first published in the *North American Review* (April 1868), and later included in *Chapters of Erie and Other Essays*, a collection of work by C. F. Adams, Jr., and Henry Adams, published in 1871.

Few themes of modern history are as compelling or obscure as the impact of mechanization upon consciousness. Although historians confidently describe many kinds of change effected by scientific technology, they discuss its influence upon the mind with manifest diffidence.[2] Of course they do take it for granted that our inventions have changed the way we think and feel. The telescope and microscope, they say, transformed man's very notion of his place in the cosmos. But they seldom make assertions of like clarity and force about the influence of industrial technology upon ideas. On this subject the standard histories of the United States offer little that is not painfully vague and banal; they tell us next to nothing about the ways in which our incomparable technology has affected prevailing attitudes and values. I propose that we can and should learn more about the effect of mechanization upon the minds of Americans. First, however, let us consider some of the reasons for our present ignorance of the subject. Why do we know so little about the impact of technological progress upon the collective consciousness?

The first reason is the exceptional difficulty of isolating for study this kind of "event" — the coming together of a technological innovation and a change in the realm of general attitudes and ideas. More than most historical problems, it seems, this one entangles us in the radical mind-matter dichotomy that has complicated the intellectual life of the West since Descartes. That is partly because the word *technology* has both physical and mental, external and internal, referents. What does it mean, for example, to speak of the *impact* made by a *machine* upon *mind*? The question is ambiguous because in this context the word *machine*, which points to a physical embodiment of technology, tacitly locates the dynamic force of change outside of consciousness; it implies that ideas are internal reflexes to objects and events "out there" in external reality. Thus a specific variant of the question — what effect did the building of the first railroads have upon American thought? — unavoidably suggests that the new machine, a tangible (material and institutional) fact, somehow *caused* certain *effects* in the general culture. In a sense it did, but the trouble with using the language of

[2] In the past few years there have been some signs of interest in the subject on the part of leading American historians. See, for instance, Perry Miller, "The Responsibility of Mind in a Civilization of Machines," *American Scholar*, XXXI (Winter, 1961); and Oscar Handlin, "Man and Magic: Encounters with the Machine," *American Scholar*, XXXIII (Summer, 1964). The establishment, in 1959, of a journal devoted to the subject (*Technology and Culture*) is perhaps the most convincing evidence of scholarly interest in the field.

causality in historical discourse is clearly evident here: it attaches to the event the character of a one-way relation between technology and culture, thereby obscuring the fact that the building of the American railroads can only be understood as a consequence of certain cultural (mental or internal) conditions, that is to say, the presence of a particular body of skills, incentives, and values in the collective consciousness. The word technology embraces both the material and mental factors, hence the railroad system is at once a product of the general culture and an influence upon it. The two stand in a reciprocal relation to each other: each in some part shapes the other. Like all historical inquiries, however, this one must begin by arbitrarily breaking what was in fact an unbroken series of interactions. As a starting point, therefore, we shall consider a period of roughly thirty years after the "moment" when the first steam-powered railroad appeared in the American landscape: 1829–1859. The problem is to relate that event to subsequent changes in the culture.

But here a second reason for our present ignorance comes into view: connections of this kind are extremely difficult to establish. They are not easily documented even when, as in the case of the telescope, they were explicitly acknowledged by those living at the time, but they seem almost impossible to document when, as in most cases, they are not acknowledged. Let us suppose, for example, that we have reason to believe a significant change in the way Americans defined their national goals coincided with (and seemed to follow from) the building of the first railroad. If this causal relationship was not recognized at the time, how are we to establish its existence? The only past we can know, after all, is the recorded past. How, then, are we to locate a nexus between action and thought unacknowledged by the actors? Problems of this kind make it imperative for historians, no less than literary critics, to devise more reliable means of access to indirect, covert, or unconscious levels of meaning in the written record of the past. (Historians, incidentally, have made surprisingly little use of the new and delicate techniques of reading in depth developed by literary criticism in recent years.) Granted that the difficulty of establishing such connections is great, it is by no means insurmountable.

This brings me to the third, final, and in many ways most important reason for our ignorance of the impact of technology upon mind. I refer to the current preoccupation of historians, to put it much too simply for the moment, with themes of power. Like most scholarship nowadays, historical writing is highly colored by an awareness of the part the United States now plays in the world. The cold war and the deployment of nuclear weapons are pervasive influences. These concerns are understandable and perhaps desirable, but they do tend to narrow the focus of scholarship; thus we have come increasingly to equate the history of the American people with the history of emergent national power. When seen from this vantage, the basis for the special affinity between technology and America seems obvious and uncomplicated. The stock explanation for this mutual attraction, which was first used in the 1840's, goes something like this: building a new society accentuates the demand for labor-saving devices, particularly when land is plentiful and labor is scarce; then, too, the absence (or relative weakness) of long-established institutions reduces the resistance to change that is to be expected in older, tradition-bound societies. Under the circumstances, and quite apart from the specific character of the culture imported from Europe,[3] it is hardly surprising that Americans have been singularly hospitable to invention. They have imported, adapted, and invented new machinery whenever it has promised to save their labor or increase their wealth, their comfort, or their general well-being. And the success of the industrial enterprise, in turn, has intensified their commitment to an ideal of progress that entails the most rapid possible rate of technological innovation.

In this view the relation between Americans and technology rests almost exclusively upon evidence of one kind of response to technical advances: the express opinions of well-defined and, more often than not, organized economic groups. It implies that the motive governing American attitudes toward change has been a more or less rational calculation of predictable economic advantage. And disadvantage: the same quantitative logic explains the

---

[3] Some cultural historians, notably the late Perry Miller, have stressed the primary importance of the subjective, or ideological, factor: the system of meaning and value brought by Europeans to the New World. Thus the Puritans, in Miller's view, were a self-selected élite with a particularly utilitarian bent; the germs of a dynamic, capitalist industrialism were latent in Puritanism from the beginning. In the main, however, American historians have emphasized the external, or environmental, factor: the stimulus imparted to technical innovation by the special conditions of life, geographical and institutional, in a virgin land. The "frontier hypothesis" of Frederick Jackson Turner is of course the *locus classicus* of this point of view. The difference between Miller's and Turner's interpretation of American development is roughly analogous to the difference between Weber's and Tawney's interpretation of the emergence of modern capitalism. But the consensus nowadays is to reject an either-or dichotomy in favor of some conception of the reciprocal play of consciousness and environment.

rare instances of organized resistance to innovation. (Item: the shipping interests opposed the protective tariff during the early years of the Republic for fear that the nascent factory system would hurt the carrying trade.) And the same simple, common-sense theory of motives (it nicely matches the dominant system of value in an open, expanding, capitalist society) will serve to account for the behavior of Americans who, while not anticipating a direct gain or loss, reacted to prospective innovations according to their presumed effect upon the prosperity of a city, state, or region. (Item: people living near the Erie Canal opposed the building of rival railroads.) When historians tacitly endorse this explanation, of course, they do not thereby deny the existence of other attitudes toward technological progress — attitudes not attributable to the manifest economic interests of any particular group. But they generally discount the historical significance of such attitudes on the pragmatic ground that they seldom issued in action, hence seldom had "practical consequences."

Given the historian's major theme, in other words, this basically economic interpretation of the American response to technological progress makes perfectly good sense. (Relevance to the theme of emergent national power is the tacit gauge of historical significance; it determines the historian's selection from the available data.) It is straightforward and simple, and it comports with the balance of the known facts. For it must be admitted that attitudes toward the machine other than those held by groups with a patent, material interest (pro or con) seldom have affected the main course of national development. And yet there is a sense in which the very unassailability of this interpretation has made it an impediment to knowledge. Because it commands so much assent, that is, we have slighted other, important though seemingly peripheral consequences of what Karl Polanyi called "the great transformation." We know too little about the impact of technology upon general (noneconomic) ideas, upon aesthetic experience, upon our direct sensory apprehension of the natural environment, upon mental health — and one could go on. Above all, we know too little about the way technology has affected the dominant structure of meaning and value. We know that mechanization has transformed the texture of our interior lives in ways that cannot possibly be reflected in the express opinions of economic groups. As an illustration, let us consider the possible relation between a change in the popular idea of America's national goals and the building of the railroad system.

The Railroad as a Cultural Symbol

Not many inventions have aroused excitement like that which accompanied the construction of the American railroads. Almost as soon as the new machine appeared, in 1829, its economic import became apparent. By accelerating the pace of the westward movement and creating vast new markets for manufactured goods, the railroad quickly stimulated the over-all process of industrialization. It confirmed the predictions of men like Tench Coxe and Alexander Hamilton, who had foreseen (in the 1790's) that conditions once universally regarded as obstacles to America's economic growth would prove to be stimulants.[4] Thus the abundance of land and the scarcity of labor, far from perpetuating the "underdeveloped" character of the economy, now increased the demand for labor-saving machinery and, in general, quickened the rate of economic growth. Between 1830 and 1860 the nation put down more than thirty thousand miles of railroad track, pivot of the transportation revolution without which industrialization under American conditions would have been impossible.[5] Today the early railroad building years may be regarded as the period of the "take-off," to use W. W. Rostow's term for what he defines as the "great watershed in the life of modern societies" when the old blocks and resistances to steady development are overcome and the forces of economic progress expand and come to dominate the society.[6]

In addition to its effect upon the economy, the railroad had an immense impact upon the public imagination. Within a few years the locomotive (variously known as "iron horse," "fire-Titan," etc.) became a kind of national obsession.[7] Its attributes were such that it seemed the very embodiment of the Age of Steam: fire, iron, smoke, noise, motion, speed, power. Newspapers and magazines of the period were filled with accounts of railroad projects, railroad

[4] Here, as in several later passages, I am summarizing an argument more fully developed and documented elsewhere. See Leo Marx, *The Machine in the Garden: Technology and the Pastoral Ideal in America* (New York: Oxford University Press, 1964), pp. 150–169.

[5] George Rogers Taylor, *The Transportation Revolution: 1815–1860* (New York: Holt, Rinehart and Winston, 1951). [Editor's Note: The chapters by Cootner and Fogel included in this book, however, attempt to cast doubt on this assertion; it is an interesting debate.]

[6] W. W. Rostow, *The Stages of Economic Growth: A Non-Communist Manifesto* (Cambridge: Cambridge University Press, 1960), pp. 7–9.

[7] "Iron horse," incidentally, is an Americanism, a fact which seems to support the notion that mechanization under American conditions had a peculiarly intense impact upon consciousness. See *A Dictionary of Americanism*, ed. Mitford M. Mathews (Chicago: University of Chicago Press, 1951).

speed, railroad accidents, railroad profits — the lore appearing in songs, poems, political speeches and stories, both factual and fictional. Here is the way Ralph Waldo Emerson describes the frenzy in 1848:

> The Railroads is [*sic*] the only sure topic for conversation in these days. That is the only one which interests farmers, merchants, boys, women, saints, philosophers, and fools. . . .
>
> The Railroad is that work of art which agitates and drives mad the whole people; as music, sculpture, and pictures have done on their great days respectively.[8]

During the 1840's, as Emerson testifies, the image of the railroad became what might be called a cultural symbol. Although it is impossible to draw a sharp line between image and symbol as I am using the terms, it is possible to describe a continuum to which they belong. By *image* I refer to a statement or verbal reference which conveys little more than a sense perception. To the degree, however, that such *images* are made to carry larger burdens of implication (thought, or feeling, or both) they approach the status of *symbols*. We may speak of the railroad as a *cultural symbol,* then, because it served to convey special meanings — beyond the level of simple reference — to a large number of those who shared the culture. The concept of a symbol that is a property of the culture generally is indispensable if we hope to understand the secondary, that is to say, indirect or even covert meanings that Americans attach to technological innovations. In the case of the railroad symbol, the primary meanings are obvious enough. Only a casual knowledge of the period is necessary to recognize that the image of the railroad was widely accepted as an embodiment of the "spirit of the times" — the Age of Steam; it was taken to represent man's newly acquired power over nature, and the idea of history as a record of virtually inevitable improvement or, in a word, progress. These primary meanings of the symbol are perfectly compatible with, and indeed may be said to confirm, the view that the dominant American response to this invention coincided with essentially economic motives. But the question here is whether there were other, second-order reactions which cannot be attributed to the mere calculation of economic advantages. To answer this question it is necessary to examine the composite image of the railroad as if it were a literary symbol. To get at the full significance of the

[8] *Journals of Ralph Waldo Emerson,* eds. Edward Waldo Emerson and Waldo Emerson Forbes (Boston: Houghton Mifflin, 1909–1914), Vol. VII, p. 504.

response, in other words, we must pay close attention to the way this symbol fits into conventional patterns of imagery, and to the connotations of thought and feeling — the overtones — which surround it.[9]

The study begins with a thorough, if not "scientific," [10] survey of the recorded responses to the new machine from all possible sources — newspapers, magazines, travel reports, political speeches, letters, drawings, diaries, sermons, and imaginative writing both popular and serious. So far as overt meanings are concerned, we find that these images of the railroad fall into an obvious pattern: a spectrum extending from unqualified approval at one pole to total repudiation at the other. However, when the images which convey little more than a sense perception are separated from those which function as symbols, we are struck at once by the frequency with which the latter are yoked to images of landscape. So striking indeed is this recurrent linkage that one is tempted to take the prominence of the setting in which the machine is placed as a gauge of its probable symbolic content. When the feelings of nineteenth-century Americans were aroused by the railroad, it seems, the affecting object was not so much the machine itself, but rather the spectacle of the machine in the natural landscape. Again and again the railroad is depicted in a terrain that is either wild or, if cultivated, rural. To appreciate the import of these images of the machine in the landscape, accordingly, it is necessary to understand the special significance of the landscape in the American imagination.

Landscape and the American Myth

Between the time the Republic was founded, in 1789, and the onset of rapid industrialization (the take-off), in the 1840's, the leading emblem of America's national aspirations was the image of an idealized rural countryside. No doubt images of landscape

[9] The approach I am describing has been used to advantage by several historians of American culture. See, for instance, Henry Nash Smith, *Virgin Land: The American West as Symbol and Myth* (Cambridge: Harvard University Press, 1950); and John William Ward, *Andrew Jackson: Symbol for an Age* (New York: Oxford University Press, 1955). For a small-scale example of the method, see Ward's essay, "The Meaning of Lindbergh's Flight," *Studies in American Culture*, eds. Joseph J. Kwiat and Mary C. Turpie (Minneapolis: University of Minnesota Press, 1960), pp. 27–41.

[10] This is not the place to discuss the merits of "scientific sampling." Suffice it to say that I have found it too mechanical and cumbersome for use in this study.

figure prominently in the national self-consciousness of every people, but the unusual circumstances of life in America invested them with a special significance. Here for the first time, said George Perkins Marsh in 1848, the full energies of "advanced . . . civilization . . . were brought to bear . . . on a desert continent."[11] The future nation was depicted as a society of the middle landscape — "middle" in the sense that it represented an ideal compromise between art and nature, between an aggressive code of technical progress and a harmonious accommodation to the nonhuman order. This society was to be located (literally and symbolically) midway between the overdeveloped nations of Europe and the primitive tribal communities of the western frontier. Neither a crowded city nor a hideous wilderness, the goal was a well-ordered green land of villages and farms — a garden enlarged to continental size and dedicated to the pursuit of happiness.

The Jeffersonian conception of a technically advanced yet rural society as the new garden of the world represented America's forthcoming redemption from history; it was to be the final chapter of the myth of the new beginning. According to the "American" myth,[12] which had emerged during the Age of Discovery, Europeans were destined to experience a regeneration in the New World. They were to become new, better, happier men — to be reborn. In most versions the regenerative power is identified with (or specifically located in) the natural terrain. Access to undefiled nature, that is, explains the special good fortune and virtue of Americans. Hence the symbolic significance of the landscape. It is a repository of latent value of all kinds: economic, political, aesthetic, religious. When Europeans have attuned themselves to the new environment, mixing civilized art with raw nature, the result is to be an ideal society of the middle landscape. As indicated by its literary and religious origins, its close resemblance to Arcadia and Eden, the mythical character of the Jeffersonian vision becomes manifest.

But though Americans enjoyed the contemplation of this vision, few were prepared to take it seriously as a guide to social policy. One of the few was Thomas Jefferson. He realized that it would be impossible to reach the goal implied by the image of the garden unless Americans were willing to repudiate some forms of economic

[11] *Address Delivered Before the Agricultural Society of Rutland County*, Rutland, Vermont, 1848.
[12] See Frederick I. Carpenter, " 'The American Myth': Paradise (To Be) Regained," *PMLA, LXXIV* (December 1959), 559–606; Marx, op. cit., pp. 141–144, 227–229.

enterprise. Hence his well-known opposition to the development of native manufactures. In *Notes on Virginia* (1785), Jefferson urged his countrymen to let their workshops remain in Europe. His argument rested upon pastoral, not agrarian, assumptions. Agrarians believed in the economic superiority of agricultural economies; Jefferson, on the other hand, was willing to defend his policy even if it proved to be an economic liability. "The loss," he said, speaking of the extra cost of transporting raw materials and manufactured goods back and forth across the Atlantic, ". . . will be made up in happiness and permanence of government." But even as he defended the pastoral ideal, Jefferson realized that his countrymen never would accept the renunciation necessary to achieve it. Their passion for commerce, power, and wealth made it unthinkable that they ever would select to remain a "backward" nation — "to stand," in Jefferson's words, "with respect to Europe, precisely on the footing of China." [13] As early as 1785, then, Jefferson realized that there was a vital difference between the society Americans were creating and their professed national goals. But it was a long while before many of his countrymen shared this insight.

The Meanings of the Railroad Symbol

Returning now to the image of the railroad, we are in a better position to understand the special fascination exercised by images of the machine in the landscape. This is not to imply, of course, that all of these images carried the same implications. For purposes of analysis they may be divided into three groups.

The triumphant machine. This is the most popular symbol, and the one, clearly, which reflects the dominant attitude.[14] The iron horse is depicted annihilating time and space, conquering all obstacles, hurtling across the continent. This machine is a Promethean earth shaker and a fire breather, and it tramples over hills, leaps

---

[13] Letter to van Hogendorp, October 13, 1785, *The Papers of Thomas Jefferson*, ed. Julian P. Boyd and Others (Princeton: Princeton University Press, 1952), Vol. VIII, pp. 631–634.

[14] The measure of popularity and dominance is not, of course, merely quantitative. To gauge the significance of these images it is necessary to take into account the shape, or structure, of institutions. In the case of images drawn from periodical literature, for example, it is necessary to discriminate between those appearing in journals which appeal to dissident, fringe elements in the society, and those appearing in journals which appeal to majority groups and to the various business and professional élites. See Marx, *op. cit.*, p. 219n.

rivers, and breaks down the gates of mountains. Usually the setting is an embodiment of a nature that is resistant, if not hostile, to man: infertile, wild, waste nature. The triumphant railroad suggests that man's dependence upon the organic is approaching an end. The picture of a locomotive conquering the vast spaces of North America makes palpable the progressive idea of history; it tends to carry strong millennial overtones. So far as this symbol may be said to derive from a serious intellectual tradition, it is that version of the Enlightenment popularly known as the "mechanical philosophy." (It implies an obscure equivalence between the mechanical perfection of the universe, as conceived in Newtonian physics, and the utopian promise of man-made machines.) But much of the meaning of the symbol is conveyed by the evocative quality of the language in which it ordinarily appears, that is to say, by attitudes and tones implicit in an ebullient rhetoric which may be called the "rhetoric of the technological sublime." [15]

The ambiguous machine. This symbol embodies (or accompanies) an overt attitude toward the railroad almost identical with the one just described. So far as the writer who invokes it makes explicit his judgment of the new machine, it is unmistakably affirmative. He is enthusiastic about the economic and social benefits promised by the railroad; he is proud of the tacit "victory" over nature, and yet he also invokes subordinate images which convey a good deal less than full confidence in the benign influence of this innovation. Often he likens the railroad to a dragon or some other monster; it breathes (sometimes vomits) fire and smoke; it heedlessly consumes natural resources; it frightens men and animals; it threatens to explode at any moment and, all in all, this symbol arouses feelings of doubt, anxiety, and disgust which cannot easily be reconciled with its ostensible beneficence. Although the overt ideas which accompany this ambiguous symbol seem to be derived from the same attenuated Enlightenment viewpoint expressed by the triumphant machine, the negative overtones hint the presence (in

---

[15] The term "technological sublime" is meant to suggest that in this period writers were imputing to machines qualities formerly reserved for natural, and often quasi-divine, objects of exalted power and grandeur. As early as 1832 a writer made this transfer of feeling quite explicit. "Alpine scenery and an embattled ocean," he said, "deepen contemplation, and give their own sublimity to the conceptions of beholders. The same will be true of our system of Railroads. Its vastness and magnificence will prove communicable, and add to the standard of the intellect of our country." Quoted in Marx, *op. cit.*, p. 195. For a more complete survey of the "rhetoric of the technological sublime," see pp. 192–206 in the same work.

an unassimilated form) of ideas most fully represented by the third category.¹⁶

The menacing machine. This symbol expresses a catastrophic view of industrialization. Here the railroad is depicted as an inhuman, hostile force, and the nature over which it triumphs is man's appointed home — the benign, green, fertile nature of the pastoral convention (and the Twenty-third Psalm). Given the physical attributes of locomotives — especially fire and smoke — the machine is easily invested with Satanic or Faustian overtones. Its speed and motion, moreover, suggest the monstrous animation of Frankenstein's robot. Faster, stronger, and more obedient than men or animals, the railroad evokes frightening thoughts of man's dispensability. It prefigures automation. This symbol appeals chiefly to those who are radically disaffected: apologists for Southern slavery, and small, ineffectual groups of socialists, transcendentalists, and sectarians. It is a favorite of the romantic movement, or counter-Enlightenment, and so far as it embraces philosophic ideas, these may be traced, via Wordsworth and Carlyle, to Germany, and to Schiller, Hegel, and Marx. The central idea here is the concept of alienation: the notion that scientific empiricism, particularly as employed in a capitalist society, separates the world of fact from the world of value and is therefore a dehumanizing force. It estranges man from nature and himself, all of which may be represented by the symbol of the railroad ravaging the countryside.[17]

The Effect of the Railroad Upon National Goals

What change in the popular conception of America's national goals is represented by these three symbols of the railroad in the American landscape? Taken by itself, the symbol of the triumphant machine seems perfectly compatible with the pastoral idea of America. Most writers who invoke it neither admit nor betray any feeling that the onset of machine power represents a shift in the direction of national life. On the contrary, they depict the conquest of the wilderness by technology as a necessary stage in the fulfillment of the Jeffersonian ideal. The steam engine is credited with

[16] The theoretical assumptions behind the idea of the ambiguous machine are more fully set forth by Bernard Bowron, Leo Marx, and Arnold Rose in "Literature and Covert Culture," *Studies in American Culture*, eds. Kwiat and Turpie, pp. 84–96; see also Marx, *The Machine in the Garden*, pp. 207–209.

[17] For a specific example, see Marx, *The Machine in the Garden*, pp. 215–220.

making the great western heartland, the Mississippi Valley, available for settlement. It is an instrument for creating a society of the "middle landscape." This is true in both a geographical and a sociological sense. The railroad literally carries civilization into the West; by the same token, its industrial concomitant, the factory system, is merely a temporary form of employment for young Americans whose ultimate ambition is to buy land. The new machine power, in other words, is a vehicle with which America will move toward a pastoral utopia. Railroads open the land to cultivation; that they also lead to the building of great cities where the garden was supposed to be is a fact rarely acknowledged by those who pay homage to the triumphant machine.

A revealing exception, however, is presented by Daniel Webster. Speaking at the opening of a railroad in New Hampshire in 1847, Webster undertakes to explain certain "prejudices" against the new road. He takes the situation of his own farm as an example. It seems that the right of way is so close to his farmhouse that the thunder of the engines and the screams of the whistle disturb the inhabitants. Besides, he says, the ugly embankment injures the look of the meadows. For a moment Webster seems to be adopting the injured tone of one who is committed to the values of the older, rural ideal: peace, repose, and happiness arising from a harmonious accommodation to "nature." But only for a moment. With a sudden change of tone — "To be serious, Gentlemen . . ." says Webster — he shifts attention to the skill and enterprise of the railroad directors, and he concludes with an orotund, Ciceronian tribute to progress in the rhetoric of the technological sublime.[18] By making an example of himself, Webster is in effect demonstrating the appropriate response to technological change. He is admitting implicitly what most writers of the "progressive" persuasion do not admit, namely, that Americans must discard the goals implicit in the pastoral image of landscape if they are to enjoy the blessings of technological progress.

That this fact was beginning to penetrate the collective consciousness is more clearly indicated by the two "minority" symbols of the railroad. In the symbol of the menacing machine the threat to the pastoral ideal was explicit. As one writer put it in 1847 (the same year as Webster's speech) after a trip through Vermont, the building of railroads means that the "beautiful pastoral life of the inhabitants will give place to oppressive factory life — quiet, rural pursuits will be absorbed in the din, conflict and degradation of

[18] See *ibid.*, pp. 209–214.

manufacturing and mechanical business. . . ."[19] The ambiguous symbol of the machine, on the other hand, reflects an unresolved conflict. The writer who invokes this symbol may be consciously, rationally committed to everything that the new technology represents, but the images he uses betray a degree of covert anxiety and uncertainty which suggests an unwillingness to relinquish the values identified with the pastoral landscape. Hence both of these symbols, the ambiguous machine and the menacing machine, reflect an awareness of the contradiction between industrialization and the older, Jeffersonian conception of America's national goals.

Summary

Before the "take-off" and the building of the railroads, Americans tended to define their nation's purpose in a pastoral idiom. The ideal society to which they aspired, or *said* that they aspired, was a harmonious society represented by the image of an improved, ordered, rural landscape. The purpose was to facilitate the "pursuit of happiness," that is, to make available to all citizens a maximum of independence, rationality, and total satisfaction. What mattered most was the over-all quality of life. The dominant goals of this good society were to be peace, harmony, and economic sufficiency — not productivity, wealth, and power. But the contradiction between this professed, pastoral ideal and the behavior of Americans, which Jefferson recognized as early as 1785, did not become apparent until the era of the "take-off." The machine's sudden appearance in the native landscape helped to make the contradiction inescapable. One of the second-order consequences of industrialization, accordingly, was to expose the untenability of the pastoral ideal, and to make clear the nation's commitment to the most rapid possible rate of economic and technological development. This newly avowed purpose generated in turn a kind of technological determinism. It encouraged the view that the single most important variable in the historical process is changing technology. "As I understand it," wrote Henry Adams to his brother, Brooks, in 1903, "the whole social, political and economical problem is the resultant of the mechanical development of power."[20] By 1933, as exemplified by the Chicago Century of Progress Exposition, this form of modern fatalism had

[19] John Orvis, "Trip to Vermont," *Harbinger*, V (July 1857), 50–52, quoted in Marx, *The Machine in the Garden*, pp. 215–217.
[20] J. C. Levenson, *The Mind and Art of Henry Adams* (Boston: Houghton Mifflin, 1957), p. 300.

become a popular creed. In the Hall of Science, for example, there was a large sculpture featuring a nearly life-sized man and woman, hands outstretched, groping, and between them stood a mammoth angular robot almost twice their size, its metallic arm thrown "reassuringly around each." The meaning of this icon was made explicit by the official motto, as recorded in the *Guidebook of the Fair*: "Science Finds — Industry Applies — Man Conforms." [21]

Popular responses to technological innovations cannot be understood adequately as the mere result of calculations of economic advantage. If this may be said of early nineteenth-century America, when the economic import of such inventions as the railroad was obvious to a large part of the population, it is more true of our time, when it has become increasingly difficult to predict the probable economic consequences of specific advances in technology. Today even the experts offer diametrically opposed opinions about the results of automation in many spheres of the economy. Today, moreover, the technological frontier is relatively remote from the immediate economic concerns of most Americans. Thus the fraction of the public directly influenced by the mammoth space effort is minuscule when compared with the fraction directly influenced by the introduction of the railroad or the automobile. To understand popular attitudes toward the space effort, accordingly, it is necessary to go far beyond explicit, rational calculations of profit and loss. It is necessary to discover how collective images of space exploration comport with existing modes of belief.

[21] Lowell Tozer, "A Century of Progress, 1833–1933: Technology's Triumph Over Man," *American Quarterly*, IV (1952), 78–81.

Index

Aaron, Daniel, 84
Adams, Brooks, 215
Adams, Charles Francis, Jr., 149, 202
Adams, Henry, 215
Adams, Wirt, Jr., 192, 193, 195, 197
Administration
 see Business administration
Agassiz, Louis, 64
Age of Discovery, 8, 12, 210
Age of Steam, 207, 208
Agricultural land, availability of, 75–81, 164
Aircraft, 127
Alliances, railroad, 144–147
American Railroad Journal, 139, 161
American Railroad Journal and Mechanic's Magazine, 133
American Society of Civil Engineers, 69
American Steamship Company, 150
Analogy, historical, 1, 2, 31, 46, 72; concept of, 2–18; *Controls for Outer Space and the Antarctic Analogy,* by Jessup and Taubenfeld, 3; in innovative industries and administration, 160–162; some generalizations, 34–36
Analogy, logical, 5–7
Arendt, Hannah, 43
Arnold, Thomas, 32
Arthur, Chester A., 174
Astor family, 157
Atchison, Topeka and Santa Fe Railroad, 69, 151
Atherton, Lewis E., 177
Atlantic and Great Western Railroad, 144, 146
Atlantic Monthly, 139
Atomic bomb, 114

Atomic reactor, 114

Baltimore Belt Railroad Company, 69, 70
Baltimore & Ohio Railroad, 56, 58, 59, 62, 69, 128, 132, 134–154 *passim,* 160
Barber, G. Putnam, 164n, 165
Barnett, H. G., 178
Barnett, S. A., 40
Bell, Daniel, 49
Belmont family, 157
Belvidere and Delaware Railroad, 69
Bergaust, Erik, 39
Berrill, N. J., 46
Bessemer, Henry, 92
Bessemer steel, 88–95, 105
Bliss, George, 160
Bloomfield, Lincoln P., 7
Boston and Lowell Railroad, 128, 132
Boston and Providence Railroad, 128, 132, 175
Boston and Worcester Railroad, 132
Boston Daily Advertiser, 132
Boston, New York and Erie Railroad, 149
Brandfon, Robert L., 24, 27–31
Bridge building, 20, 21, 62, 63, 65, 69, 140
Brindley, James, 59n
Brooklyn Polytechnic Institute, 170
Brooklyn subway, 69, 70
Brotherhood of Locomotive Engineers, 159
Brown, Gordon, 20n
Brunel, Isambard Kingdom, 34
Buckhout, Isaac C., 69
Bureaucracy, 24, 25, 29, 143–162, 170
Burlington Railroad
 see Chicago, Burlington & Quincy

INDEX

Business administration, 21, 127–162, 170, 171; evolution of, 127–143; implications of new forms, 158–160; methods revolutionized, 134, 135
   see also Managerial aspects

California, railroad commission, 176
Camden and Amboy Railroad, 128, 173
Canals, 20, 22, 54, 55, 57, 61, 72, 76, 79–82, 97, 102, 108, 109, 111, 112, 123, 129, 130, 170, 172
Cape Kennedy, 49
Carlyle, Thomas, 32, 213
Cartesian philosophy, 3
Cavour, Count, 30
Carruth, Nathan, 133
Castle Garden Labor Bureau, 167
Central Railroad of New Jersey, 149
Champagne Canal, 172
Chandler, Alfred D., Jr., 21–29 *passim*, 34, 35, 48, 49n, 121, 170
Chesapeake and Ohio Railroad, 158, 188
Chester and Holyroad Railway, 63n
Chicago and Northwestern Railroad, 148, 151, 152
Chicago Board of Trade, 176
Chicago, Burlington & Quincy Railroad, 141, 148, 151, 152, 154, 159
Chicago Century of Progress Exposition, 215
Chicago, Milwaukee and St. Paul Railroad, 151
Chicago, Rock Island and Pacific Railroad, 148, 151, 152, 154
Chicago, St. Louis and New Orleans Railroad, 185, 190
Cincinnati, Hamilton and Dayton Railroad, 147
Clark, Colin G., 5
Clark, Horace, 147, 155
Clarke, J. C., 171
Cochran, Thomas C., 16, 21, 25–29, 31, 33, 50
Cohen, Morris R., 6
Columbus, Chicago and Indiana Central Railroad, 144

Commissions, canal, 172; railroad, 172, 175, 176
Competition, impact of, 143–146, 151, 152, 184
Connecticut, railroad commission, 175
Cook, Thomas, and Sons, 28
Cooper, Theodore, 69
Cooper Union, 170
Cootner, Paul H., 21–23, 25, 26, 28, 35, 47
Copernicus, Nicolaus, 43, 46, 51
Corning, Erastus, 141
Costs, 75, 77, 78, 80, 129
Coxe, Tench, 207
Crozet, C., 57
Curti, Merle, 165

Dartmouth College, 170
Darwin, Charles, 46, 51
Davis, Kingsley, 5
Day, Henry Noble, 178
Delaware and Hudson Canal, 170
Demography, 26, 163–169
Derby, Elias Hasket, 133
Descartes, René, 46, 203
Deutsch, Karl, 3
Diamond, Edwin, 39
Donne, John, 38
Douglas, Stephen, 183
Drew, Daniel, 140

Eads, James B., 20, 62, 63, 185
Eastern Trunk Line Association, 149, 150
Economic aspects, 21–23, 26, 27, 38, 47, 56, 57, 74–126, 182; importance of technical competition, 108–110; organization, 119–123; patterns, 110–112, 116; timing and magnitude, 112–116
Education, 39, 63, 64, 71, 122, 169, 170
Einstein, Albert, 43
Elevated railways, 68–71
Eliade, Mircea, 7
Elkins, Stanley, 2n
Emerson, Ralph Waldo, 31, 177, 208
Environment, man-made, 21, 38, 41, 42, 53, 60, 61, 71
Erie Canal, 55, 62, 129, 172, 206

Erie Railroad, 24, 25, 69, 95, 130, 138–140, 143, 144, 146, 148, 149, 151, 152, 158, 160
Eustis, Henry, 64
Evans, Oliver, 122

Family, impact on, 169
Fears and phobias, 32, 33, 62, 179, 213
Federal Mississippi River Commission, 187
Finances, 22, 45, 47, 121, 152, 156–158, 161, 183, 189–194 *passim*
see also Investment
Fink, Albert, 149, 152
Fish, Hamilton, 184
Fish, Stuyvesant, 157, 183–185, 196–198, 200
Fishlow, Albert, 94n, 103
Fogel, Robert William, 21–23, 35, 109, 113, 115, 116
Freedman, Toby, 38
Freud, Sigmund, 32
Frontier problems, 54–60
Fuel, demand for, 118, 119

Galileo, Galilei, 4
Galton, Douglas, 139
Garden Cities of Tomorrow, by Howard, 66
Gagarin, Major Yuri, 40
Garrett, John W., 148, 152
General Motors, 152, 156
Geophysics Corporation, 38
Georgia Railroad, 140
Gibbs, Wolcott, 64
Gilfillan, S. C., 12, 13
Gillespie, W. M., 64, 65
Goldsen, Joseph M., 3, 38n
Gould, Jay, 25, 146–156 *passim*
Government assistance, 1, 25–27, 48, 127, 171–177
Government regulation, 25, 122, 123, 174–177
Grady, Henry, 186
Grand Central Station, 69, 70
Grand Trunk, Canada, 148
Grandis, Sebastiano, 62
Granger movement, 30
Grattoni, Severino, 62

Gray, Asa, 64
Great Western Railway, 34
Griswold, John N. A., 159

Hale, Nathan, 132
Hamilton, Alexander, 207
Hanger, Harry B., 70
Harding, Francis, 38
Harriman, Edward H., 157
Hartwell, England, 114
Harvard University, 64
Hawthorne, Nathaniel, 31
Healy, Kent T., 76
Hegel, G. F. W., 15, 213
Hexter, J. H., 3, 4
Hobbes, Thomas, 4
Hoosac Tunnel, 61–63
Hofstadter, Richard, 83
Hoselitz, Bert F., 12n
Howard, Ebenezer, 66
Hughes, Thomas, 19, 21–23, 41, 42, 44, 47, 50
Hume, David, 4
Hunt, Robert S., 172, 174
Hunter, Louis, 120
Huntington, Collis P., 188, 189
Huntsville base, NASA, 49, 50
Huskisson, William, 32

Illinois Central Railroad, 24, 28, 29, 135, 141, 152, 156–158, 161, 171; in Mississippi, 182–201
Imaginative impact
see Intellectual, imaginative, and psychological impact
Independent Subway System of New York City, 70
Indiana Central Railroad, 146, 147, 150
Industrial Revolution, 4, 7, 10, 12, 19n
Innovation and invention, impact of, 23, 24, 47, 62, 63, 74, 105; on business administration, 127–162; economic, 107–126
Innovators, in business administration, 127–162
Intellectual, imaginative, and psychological impact, 27, 28, 31–33, 36, 41, 47, 50, 51, 163, 202–216
Interstate Commerce Commission, 122

INDEX

Invention, social, 12, 13, 14
Investment, 22, 24, 35, 74, 111, 113, 131, 171
  see also Finances
Iowa Pool, 148
Iron and steel manufactures, 83–92 passim, 111, 116–118, 120, 123
Iron ore and coal, availability of, 81–83, 118, 119

Jackson, Patrick Tracy, 132
James River Valley, survey of, 57
Jastrow, Robert, 38, 46
Jeans, J. S., 92
Jefferson, Thomas, 31, 210, 211, 213, 215
Jenks, L. H., 23n
Jervis, John B., 135
Jessup, Philip C., 3
Joint Executive Committee, 149
Joy, James, 152

Katte, Walter, 69
Kelly, William, 92
Keep, Albert, 152
Kennedy, John F., 8, 40, 41
Kilby Tunnel, 62
Kirkland, Edward C., 174, 175, 177
Kropotkin, Peter, 66

Labor unions, 158–160
Lackawanna Railroad, 149
Lake Shore and Michigan Southern Railroad, 147, 155
Landscape and the American myth, 209–211
Languedoc Canal, 61
Lardner, Dionysius, 57
Latrobe, Benjamin H., 135, 137, 138, 151
Laurie, James, 69
Lawrence, Amos, 170
Lawrence Scientific School, Harvard, 64, 170
Ledyard, Henry B., 176
Lee, Everett, 163
Legal aspects, 152, 153
Lehigh Canal, 170
Leicester & Swannington Railway, 34
Lincoln, Levi, 172

Lipset, Seymour M., 16
List, Friedrich, 29
Liverpool and Manchester Railway, 19, 32, 128
Lobbying, 172–174
Louisville & Nashville Railroad, 152
Louisville, New Orleans and Texas Railroad, 188–198 passim
Lowe, Abraham, 133
Lowenthal, David, 33n
Lumber products, 86, 87

McAlpine, William J., 69
Macaulay, Thomas B., 30, 37
McCallum, Daniel C., 24, 134, 135, 138–140, 151
McClellan, George B., 135
MacIver, Robert M., 13
McKitrick, Eric, 2n
McLane, Louis, 135, 137
McLaurin, Anselm J., 198
Maine, railroad commission, 175
Mallory, George, 40
Managerial aspects, 24–35 passim, 47–49, 121–123
  see also Business administration
Manchester, England, 66
Manufactured products, market for, ante bellum period, 83–87; post-Civil War era, 87–92
Marietta and Cincinnati Railroad, 145
Marsh, George Perkins, 210
Marshall Space Flight Center, 200
Marx, Karl, 15, 213
Marx, Leo, 28, 31–33, 35, 50
Massachusetts Institute of Technology, 170
Massachusetts Railroad Commission, 149, 175
Mazlish, Bruce, 53n
Medawar, P. B., 45
Michaud base, NASA, 50, 200
Michigan Central Railroad, 135, 141, 147, 148, 152, 159, 172, 176
Michigan Southern Railroad, 146, 147
Michigan, University of, 64, 170
Military aspects, 74, 135
Miller, George P., 1
Miller, Perry, 205n
Miller, William, 84

Mississippi, railroad monopoly in, 182–201
Mississippi Test Facility, 200
Mississippi Valley Company, 194
Mitchell, B. R., 23n
Moore, Wilbert E., 13n
Morgan, J. Pierpont, 156, 157, 162
Morris, Gouverneur, 172
Motor vehicles, 105, 106, 127, 166
Mt. Cenis Tunnel, 21, 61, 62
Mumford, Lewis, 13n, 39, 42, 43

Nagel, Ernest, 6
National Aeronautics and Space Act of 1958, 40
National Aeronautics and Space Administration, 200; bases, 49, 50, 200
National Conference and Space Exposition, Fourth (1964), 42
National goals, 213–216
New Haven, Hartford and Springfield Railway, 69
New York and Boston Rapid Transit Company, 69
New York and Erie Railroad, 128, 132, 134, 138, 141
New York and Harlem Railroad, 69
New York Central Railroad, 24, 25, 129, 138, 141–144, 147, 148, 151, 152, 158, 171, 172, 173; reorganized, 155, 156
New York Elevated Company, 69
New York, New Haven, and Hartford Railroad, 70
New York State Railroad Commissioners, 139, 175
Newell, Homer E., 38, 46
Newman, Cardinal, 30
Northern Pacific Railroad, 174

Oak Ridge, Tennessee, 49
Ogburn, William F., 13
Ohio and Mississippi Railroad, 147
Olschki, Leonardo, 4
Organization of the Service of the Baltimore & Ohio Railroad, 136
Osborne, William H., 152, 184, 185

Panhandle Railroad, 144, 152, 153
Pascal, Blaise, 23
Pennsylvania Company, 152, 153
Pennsylvania Railroad, 24, 25, 69, 70, 129, 130, 134, 135, 140–161 *passim*, 171, 172
Perkins, Charles E., 152, 176
Philadelphia, Wilmington and Baltimore Railroad, 128
Philosophic impact, 41–44
Pierce, Harry H., 178
Pipes, Richard, 3
Pittsburgh Daily Gazette and Advertiser, 177
Pittsburgh, Fort Wayne and Chicago Railroad, 144–147, 150
Polanyi, Karl, 206
Political impact, 21, 27–32, 36, 47, 50, 56, 57, 74, 182–201
Political Man, by Lipset, 16, 17
Poor, Henry Varnum, 139
Pope, Alexander, 33
Population, redistribution, 76, 164
see also Demography
Practical Treatise on Rail-Roads and Carriages . . . , by Tredgold, 60
Price, Derek, 48
Progressivism, 30
Prophecies, 36–39
Pruyn, John V. L., 141
Psychological impact
see Intellectual, imaginative, and psychological impact
Purposes, statements of, 39–41

Quality production, impact on, 120, 121
Quincy Railway, 58, 59

Railroad Age, 12
Railroad Gazette, 167
Railway Economy: A Treatise on the New Art of Transport, by Lardner, 57
Railways Terminating in London . . . , by Rea, 69
Rea, Samuel, 69
Reade, Winwood, 37
Reading Railroad, 158
Reagan, J. F., 176

Rensselaer Polytechnic Institute, 64, 169, 170
Resources, effect of availability of, 75–83, 104; agricultural land, 75–81; iron ore and coal, 81–83
Rhode Island Railroad Commission, 175
Richmond, Dean, 141
Roads
    see Turnpikes
Robinson, A. P., 67
Rock Island Railroad
    see Chicago, Rock Island and Pacific Railroad
Rosenstein-Rodan, Paul, 115
Rostow, W. W., 23

Saint-Simonians, 30
Salmons, Charles H., 159
Salsbury, Stephen, 21, 23–26, 28, 29, 34, 35, 48, 49n, 121, 170
Santa Fe Railroad
    see Atchison, Topeka and Santa Fe Railroad
Schiller, J. C. F. von, 213
Scott, Thomas, 152
Secondary industries, effect of railroad expansion, 116–119
Sheffield Science School, Yale, 64, 170
Siemens brothers, 93
Sloan, Alfred, 156
Smith, Adam, 15
Social Change, by Moore, 13
Social Change, by Ogburn, 13
Social impact, 25–28, 46–50; demographic effects, 163–169; institutional effects, 163, 169–177; social-psychological effects, 27, 163, 177–181
Social invention, 13, 18–52; as a form of historical analogy, 12
Social savings, 96–103
Society; its Structure and Changes, by MacIver, 13
Sociology of Invention, The, by Gilfillan, 12
Sommeiller, Germain, 62
Sooysmith, Charles, 69
South Carolina, railroads in, 132
Southern Pacific Railroad, 154, 188

Southern Railroad, 195, 196
Southern railroads, 158
Space Age, 8, 12, 37, 39, 42
Space program, analogies in, 3, 47–51, 71–73, 106, 107, 125, 126, 160–162, 200, 201, 216; philosophical impact, 41–44; scientific consequences, 45–47; social invention in, 18–52; technological frontier, 21, 27, 44, 45, 53–73
    see also Business administration, Economic aspects, Social impact, Innovation and invention, Political impact
Space Revolution, 7
Spengler, Oswald, 2n
Stages of Economic Growth, The, by Rostow, 23
Standard Oil of New Jersey, 152
Steel
    see Iron and steel manufactures
Stephenson, George, 7, 34, 69n, 112, 122
Stephenson, Robert, 34, 63n, 69n, 122, 128
Steubenville and Indiana Railroad, 145
Stevens, John, 58, 60
Street railways, 67–69
Subways and elevated railways, 68–71
Symbols
    cultural, 207–209
    railroad, 211–213

Taubenfeld, Howard J., 3
Tawney, R. H., 205n
Technics and Civilization, by Mumford, 13n
Techniques, acquired, 61–65; transfer of, 65–70
Technological frontier, 21, 27, 44, 45, 53–73
Technology, 12, 13, 19–21, 47, 105; imaginative aspects, 202–216; innovations in, effects of railroads, 92–96, 107–126
Telegraph, 95, 96
Teller, Edward, 43n
Temin, Peter, 94
Textile mills, 111, 122, 130, 132

Thomson, J. Edgar, 135, 140, 145–161 *passim*
Thomson, James, 31, 32, 51
Thoreau, Henry David, 31
Time zones, 180
Tocqueville, Alexis de, 33
Toledo, Wabash and Western Railroad, 146, 149
Tracy, John F., 152
Tredgold, Thomas, 57, 60
Trevithick, Richard, 112, 122
Troy and Greenfield Railroad, 69
Tunnel construction, 61–63
Turner, Frederick Jackson, 71, 163, 205n
Turnpikes, 20, 22, 54, 72, 80, 102, 108, 109, 123, 164, 166

Underground Transit Railway of New York, 69
Union Canal, 81
Union College, 64
Union Pacific Railroad, 173
United States Military Academy, 160, 170
United States Steel Corporation, 152, 156, 162
United States Trust Company, 194
Urbanization, 21, 22, 26, 42, 66–68, 71, 119, 168

Vanderbilt, Cornelius, 142, 147–150, 152, 155
Vanderbilt, William H., 142, 147–150, 71, 119, 168
van Rensselaer, Stephen, 170
Vermont, railroad commission, 175
Vermont Central Railroad, 148

Verne, Jules, 37
Vibbard, Chauncy, 141, 142
von Braun, Wernher, 49, 200

Wabash
  see Toledo, Wabash and Western Railroad
Wagon transportation, 77, 78, 80
Waterways, 76–79, 97, 108, 123, 129, 130, 164–167, 187, 195
Watt, James, 112
Webb, James E., 3
Weber, Max, 205n
Webster, Daniel, 1, 8, 214
Wells, David A., 149
West Point
  see United States Military Academy
Western Executive Committee, 149
Western Railroad (Massachusetts), 27, 62, 128, 129, 131–135, 160, 161, 172, 173, 178
Westinghouse, George, 124
Whistler, George, 133, 160
Williams, Robin M., Jr., 180
Windom Report, 176
Wohl, R. Richard, 178
Wood, Nicholas, 56n
Worcester, Edwin D., 141
Wordsworth, William, 31, 213
Wright, J. A., 149

Yale University, 64, 170
Yazoo and Mississippi Valley Railroad, 189, 190, 192–196, 198, 199
Yazoo Mississippi Delta, 186–191, 195, 196, 199

Zelikoff, Murray, 38